Lectures on Quantum Field Theory and Functional Integration

Zbigniew Haba

Lectures on Quantum Field Theory and Functional Integration

Zbigniew Haba
Theoretical Physics
University of Wrocław
Wrocław, Poland

ISBN 978-3-031-30714-0   ISBN 978-3-031-30712-6 (eBook)
https://doi.org/10.1007/978-3-031-30712-6

This Springer imprint is published by the registered company Springer Nature Switzerland AG
The registered company address is: Gewerbestrasse 11, 6330 Cham, Switzerland

To Julia

Preface

The purpose of most textbooks on quantum field theory (QFT) is a development of a theory for a description of elementary particles. We have a modest aim: to describe methods that express quantum field theory as quantum mechanics in an infinite number of dimensions. With such an attitude we pay less attention to symmetries and kinematics. We concentrate more on integration and analysis in an infinite number of dimensions as a framework for QFT. At the same time, we wish to include the standard results of quantum field theory: the scattering amplitudes in a perturbative expansion. In the conventional formulation of QFT, the functional integral is treated as a formal tool for a quick derivation of the diagrammatic rules for perturbative calculations. It is known that the functional integral has a rigorous mathematical formulation at an imaginary time. Then, QFT has an interpretation as the classical statistical mechanics. It becomes available for numerical simulations. We pay some attention to the relation between the real and imaginary time and to the functional integration in the real time. The main omission of these lectures concerns the fermionic fields. In a sense, the fermionic fields do not fit to the functional integration scheme (without Grassmanian variables). The reader interested in particle physics after reading this introductory text should take another textbook to learn about quantization of quarks and electrons. We do not discuss the antisymmetric correlations of electron and quark fields although we spent some time in the studies of a charged particle in a quantum electromagnetic field. We are careful with mathematics but we do not insist on complete proofs rather indicating how one could do them. We also consider rather informal mathematical definitions in the way physicists use them rather than insisting on the mathematical rigor (we refer to mathematical literature for precise definitions and proofs). We use some questionable methods of QFT, e.g., the interaction picture (which is false according to the Haag theorem) and the Gell-Mann-Low formula (which exploits a doubtful oscillatory limit). These methods are correct with proper regularizations leading to the standard perturbative results of QFT. The nonperturbative aspects are approached by means of functional integration. We pay more attention than usual to the relation between operator and functional integral approaches. Some problems are simpler to solve in an operator framework but some other results are easier to derive in the functional integral formulation. For this reason, these lectures can also be of some interest for mathematicians as some functional integration formulas for random fields can be obtained in an operator framework. On the other hand, the functional integration methods and stochastic differential equations can be useful in numerical simulations in QFT. The lectures contain the main results and methods from standard QFT of scalar, electromagnetic and Yang-Mills fields both in Minkowski and Euclidean formulation. The Euclidean approach may be of interest for people interested in viewing quantum field theory as a part of classical statistical physics.

In Chap. 1 we define the main mathematical tools needed in quantum field theory which do not belong to the elementary calculus. In Chap. 2 we quantize classical field theory as a quantum mechanics in an infinite number of dimensions. In Chap. 3 the interaction is introduced. We derive the main formulas for the scattering matrix. Some supplementary methods used in contemporary quantum field theory (fields on a manifold and at finite temperature) are described in Chap. 4. On the basis of the operator methods developed in first three chapters we develop in Chap. 5 the formalism of the functional integration as equivalent to the operator framework. We apply the rigorous mathematics of Chap. 5 to some models of quantum mechanics and quantum field theory in Chap. 6. In Chap. 7 we apply the functional integration on a formal level (as it is usually treated by physicists) in order to derive the main results of perturbative quantum field theory. Chapter 8 goes beyond the standard methods of functional integration as it attempts to use the paths of the free field for the path integral. In the Hilbert space framework, the transformation of paths corresponds to a transformation from the formal Hilbert space $L^2(d\phi)$ to a mathematically well-defined $L^2(|\psi_0|^2 d\phi))$, where ψ_0 is the free field ground state. We apply this method for a description of the field on a manifold. In Chap. 9 the interaction of the quantum electromagnetic field with charged particles is discussed in the framework of non-relativistic quantum mechanics (in the way it is exploited in quantum optics). In a similar way, we treat in Chap. 10 a particle interaction with quantum gravitational waves. Chapter 11 concerns the quantization of non-Abelian gauge fields. For a particle physicist, the development of the quantum theory of gauge fields may be considered as the main motivation for using the functional integral. The lattice approximation of Chap. 12 may be considered as a rigorous (computer friendly) version of imaginary time (Euclidean) quantum field theory of scalar and gauge fields. The bibliography is not complete and does not reflect the historical achievements of the researchers in the field. It just provides the reader with a basic (and subjective) source of information on the subject under study.

These lecture notes arose from my lectures in the Institute of Theoretical Physics at Wroclaw University for students with a basic knowledge of calculus, classical physics and quantum mechanics. For this reason, my lectures begin with elementary material emphasizing student's experience with quantum mechanics and gradually approaching advanced topics. At this stage of their studies, the students are looking for problems for research. I intended to supply a broad range of topics for exploration besides the basic knowledge. I tried to develop some tools useful for solving technical problems. The exercises at the end of each chapter fill omissions in the main text or suggest to use the results for further applications.

My treatment of QFT profited a lot from my discussions and collaborations with Sergio Albeverio, Erhard Seiler and Hagen Kleinert. I am grateful for a creative scientific atmosphere in Wroclaw due to professors Jan Lopuszanski, Jerzy Lukierski and Arkadiusz Jadczyk and for their support for my interest in QFT in my early stage at Wroclaw University.

Wrocław, Poland   Zbigniew Haba

Contents

Chapter 1
Notation and Mathematical Preliminaries

Abstract In the first chapter we introduce the main mathematical supplements to the standard course of analysis which are necessary in the studies of quantum field theory (QFT). We outline these topics in a minimal extent (referring to the easily available literature): the theory of distributions, the functional differentiation (differentiation in infinite number of dimensions) and the basic notions of group theory. We discuss in more detail the Gaussian functional integral as an integral in infinite number of dimensions.

We use the convention of a summation over repeated indices. The indices of vectors in the Minkowski space M are raised with the Minkowski metric η defined by the scalar product

$$x^\mu y_\mu = x_0 y_0 - x_k x_k \equiv x_0 y_0 - \mathbf{x}\mathbf{y} \equiv \eta^{\mu\nu} x_\mu y_\nu$$

for $x, y \in M$.

The Greek indices have the range $\mu = 0, 1, 2, 3$ whereas the Latin indices $k = 1, 2, 3$. We discuss also the Euclidean metric

$$x_\mu y_\mu = x^\mu y^\mu = x_0 y_0 + x_k x_k \equiv \delta_{\mu\nu} x^\mu y^\nu$$

for $x, y \in R^4$. If the repeated indices are on the same level then this means the Euclidean metric (either four dimensional or three dimensional). For the most part of the text we set the Planck constant $\hbar = 1$ and the velocity of light $c = 1$ except of the sections where the dependence on these constants has to be exposed. The Lebesgue integration measure over M (in various dimensions) will be denoted simply as dx except of the cases when we wish to emphasize that the integration is over the four dimensional volume, then we use the notation d^4x. The three dimensional vectors are denoted by bold letters $\mathbf{x} \in R^3$. We also consider $\mathbf{n} \in Z^d$, where the vector $\mathbf{n} = (n_1, \ldots, n_d)$ has integer components. We use the notation $d\mathbf{x}$ for an integration over a three dimensional volume. The complex conjugation of $z \in C$ is denoted z^*. The Hilbert space of square integrable functions with respect to a measure μ is denoted $L^2(d\mu)$, the scalar product of $f, g \in L^2(d\mu)$ is written as $(f, g) = \int d\mu f^* g$. By $(f, g) \equiv f(g)$ we also denote the linear functional $f \in S^*$ (the dual space to

S) defined on functions $g \in S$. The Brownian motion $\mathbf{w}$ appears in various forms (beginning in $\mathbf{0}$, or in $\mathbf{x}$ or as a pinned Brownia motion beginning in $\mathbf{x}$ and ending in $\mathbf{y}$). We avoid an introduction of separate notations for each of them explaining the case at the palace where it appears. The average over the various Brownian motions is denoted either by $dW(\mathbf{w})$ or as $E[..]$. The latter notation is applied in general for any expectation value over random paths. The heuristic Feynman path integral, in accordance with tradition, is written as $\mathcal{D}\mathbf{q}$. However, when we mean a functional integral over fields ϕ we denote it like a formal Lebesgue measure $d\phi$ which requires a mathematical definition usually together with a Gaussian factor. This allows to define the integration as an integral with respect to a Gaussian measure. Mathematical objects appearing locally are explained at the place where they are used. In the text numerous constants appear as a result of calculations (or as normalization constants). We denote them usually as K or Z what does not mean that they take the same value. It would be a mess to denote them with different letters.

1.1   Generalized Functions (Distributions)

We define a generalized function $F \in S^*$ as a linear functional on a space of real functions S (for simplicity we assume here that functions are defined on R), i.e., for $\alpha, \beta \in R$

$$F(\alpha f_1 + \beta f_2) \equiv (F, \alpha f_1 + \beta f_2) = \alpha(F, f_1) + \beta(F, f_2). \qquad (1.1)$$

As S we may take the set of infinitely differentiable functions vanishing fast at infinity (see [105] for more details). $S \subset S^*$ as for $f \in S$ we can define the linear functional

$$f(g) \equiv (f, g) = \int dx f(x)g(x). \qquad (1.2)$$

As an example, the Dirac's $\delta(x - y)$ is defined as the linear functional

$$(\delta_y, f) = f(y)$$

which agrees with the formal definition $\int dx \delta(x - y) f(x) = f(y)$.
   The derivative of a distribution F is defined by the formula

$$(F', f) = -(F, f') \qquad (1.3)$$

in agreement with the integration by parts formula for $F \in S$. We still need the notion of the Fourier transform $\tilde{F}$ of the distribution F

$$(\tilde{F}, f) = (F, \tilde{f}), \qquad (1.4)$$

where

$$\tilde{f}(x) = (2\pi)^{-\frac{1}{2}} \int dp \exp(ipx) f(p). \tag{1.5}$$

It follows from Eq. (1.3) that the definition of a derivative is an extension of the derivative from $\mathcal{S}$ to $\mathcal{S}^*$. Similarly, it can be seen that the definition of the Fourier transforms (1.4) follows from the Plancherel formula in $\mathcal{S}$

$$(f, g) = (\tilde{f}, \tilde{g}). \tag{1.6}$$

As an application of Eq. (1.3) we can derive (Exercise 1.1)

$$\partial_t \theta(t) = \delta(t)$$

(where θ is the Heaviside function, $\theta(t) = 1$ for $t \geq 0$ and $\theta(t) = 0$ for $t < 0$). From Eq. (1.4)

$$\delta(x) = \frac{1}{2\pi} \int dp \exp(ipx).$$

These formulas will be frequently used in subsequent chapters.

1.2 Functional Differentiation

We define the Frechet derivative Φ' of a (non-linear) functional $\Phi(\phi)$ (where $\phi \in \mathcal{S}^*$) as an element of $\mathcal{S}^*$ defined by the formula [122] ($s \in R$)

$$(\Phi', f) = \lim_{s \to 0} s^{-1} \Big(\Phi(\phi + sf) - \Phi(\phi) \Big). \tag{1.7}$$

Here, it is assumed that the rhs of Eq. (1.7) is linear in $f \in \mathcal{S}$ and defines a linear functional on $\mathcal{S}$, so that the rhs defines an element of $\mathcal{S}^*$. We use the notation

$$(\Phi', f) = \left(\frac{\delta\Phi}{\delta\phi}, f \right) \equiv \int dx \frac{\delta\Phi}{\delta\phi(x)} f(x). \tag{1.8}$$

The virtue of such a notation comes from a recognition of some properties of the functional derivative similar to the elementary derivative. For example if Ψ is a function of Φ then

$$\frac{\delta\Psi}{\delta\phi(x)} = \Psi'(\Phi) \frac{\delta\Phi}{\delta\phi(x)}. \tag{1.9}$$

Equation (1.9) together with

$$\frac{\delta\phi(y)}{\delta\phi(x)} = \delta(x - y) \tag{1.10}$$

allows to calculate most functional derivatives appearing in these lectures.

Calculating the Frechet derivative of (the linear functional) $\frac{\delta\Psi}{\delta\phi(x)}$ once more we obtain $B(x, y) = \frac{\delta^2\Psi}{\delta\phi(x)\delta\phi(y)}$ which is a kernel of a bilinear form on $\mathcal{S}$, i.e., a kernel (see Sect. 3.4) of an operator. Continuing the differentiation n times we obtain as a result an n-linear form on $\mathcal{S}$. For explicit calculations see Exercise 1.5.

1.3   Gaussian Integration

We consider the formula of Gaussian integration ($X, J \in R^n$, M is $n \times n$ matrix such that $\Re M$ is positive definite) then

$$Z[J] = \int dX \exp\left(-X\frac{M}{2}X\right)\exp(iJX) = \det\left(\tfrac{1}{2\pi}M\right)^{-\frac{1}{2}}\exp\left(-\tfrac{1}{2}JM^{-1}J\right),$$
(1.11)

where the square root is chosen (by an analytic continuation) in such a way that $\det(\frac{1}{2\pi}M)^{-\frac{1}{2}} > 0$ when M is real and positive definite. For a real Hermitian M the formula can be shown by means of a change of variables $X \to X - M^{-1}J$. The proof for complex M is obtained by means of an analytic continuation (see [89], Appendix A). It follows from Eq. (1.11) that

$$\int dX \exp\left(-X\frac{M}{2}X\right) = \det\left(\frac{1}{2\pi}M\right)^{-\frac{1}{2}}.$$
(1.12)

We still need another result from [89] (see also [76, 154]). Let $M = -iB$ where B is a real invertible symmetric matrix then (in the sense of a Fourier transform of a distribution)

$$\int dX \exp\left(iX\frac{B}{2}X\right)\exp(-iJX) = \det\left(\tfrac{1}{2\pi}(-iB)\right)^{-\frac{1}{2}}\exp\left(-\tfrac{i}{2}JB^{-1}J\right)$$
$$= \exp\left(-\tfrac{i}{2}JB^{-1}J\right)\det\left(\tfrac{1}{2\pi}|B|\right)^{-\frac{1}{2}}\exp\left(i\tfrac{\pi}{4}ind(B)\right),$$
(1.13)

where $ind(B)$ is an integer equal the number of positive eigenvalues of B minus the number of negative eigenvalues. The formula (1.13) is derived from (1.11) as the limit of $M = \epsilon I - iB$ when $\epsilon \to 0$. The choice of the square root is determined by the analytic continuation mentioned at Eq. (1.11).

The formula (1.11) can be generalized to infinite number of dimensions. In particular, we may assume that X is a (generalized) function. We consider the (Gelfand) triple $\mathcal{S} \subset \mathcal{H} \subset \mathcal{S}^*$ where $\mathcal{H}$ is a Hilbert space. $JX \to \int dx X(x)J(x) \equiv (X, J)$, M is an invertible symmetric positively definite operator on $\mathcal{H}$ and $(J, M^{-1}J)$ in Eq. (1.11) is a bilinear form on $\mathcal{S}$.

$Z[J]$ satisfies the inequality

$$\sum_{i,k=1}^{n} c_i c_k^* Z[J_i - J_k] \geq 0, \tag{1.14}$$

where c_i, $(i = 1, \ldots, n)$ are complex numbers.

The inequality (1.14) can be shown if we apply the finite dimensional approximation (1.11) and finally take the limit when the dimension goes to infinity. However, for further purposes we prove directly that if the real symmetric bilinear form JGJ is positive definite then

$$Z[J] = \exp\left(-\frac{1}{2}JGJ\right) \tag{1.15}$$

satisfies the inequality (1.14). In fact, the inequality (1.14) for (1.15) can be expressed as

$$\sum_{i,k=1}^{n} \left(c_i \exp\left(-\frac{1}{2}J_i G J_i\right)\right)\left(c_k \exp\left(-\frac{1}{2}J_k G J_k\right)\right)^* \exp(J_i G J_k) \geq 0$$

Now, the inequality (1.14) follows from the following statement ([104], Theorem 4, Chap. III):

If the matrix $G = (G_{ik})$ is non-negative in the sense that for arbitrary complex numbers c_j

$$\sum_{i,k=1}^{n} c_i c_k^* G_{ik} \geq 0$$

then the matrix A with matrix elements $A_{ik} = \exp(\tau G_{ik})$ for any $\tau > 0$ is also non-negative, i.e.,

$$\sum_{i,k=1}^{n} c_i c_k^* A_{ik} \equiv \sum_{i,k=1}^{n} c_i c_k^* \exp(\tau G_{ik}) \geq 0. \tag{1.16}$$

The proof follows from the formula

$$\exp(\tau A) = \lim_{n\to\infty}\left(1 + \frac{\tau}{n}G\right)^n,$$

where the matrix $1 + \frac{\tau}{n}G$ is non-negative by assumption. Then, its arbitrary power is non-negative and the limit $n \to \infty$ is non-negative.

In the case of $Z[J]$ of Eq. (1.15) $G_{ik} = J_i G J_k$, $\sum_{ik} c_i c_k^* G_{ik} = (\sum_i c_i J_i) G(\sum_k c_k J_k)^* \geq 0$. The inequality (1.14) has as a consequence the theorem of Minlos-Sazonov (see [104, 172, 225] for a precise formulation). It says that if we have the functional $Z[J]$ on $\mathcal{S}$ (with some continuity properties) satisfying the inequality (1.14) then there exists a measure μ on the dual space $\mathcal{S}^*$ such that

$$Z[J] = \int d\mu(X) \exp(i(J, X)). \tag{1.17}$$

The rhs of Eq. (1.15) satisfies the inequality (1.14) hence it defines a measure (the Gaussian measure $d\mu_0$). If we multiply $d\mu_0$ by an integrable real function $\exp(-\int V)$ (where $\int V$ will be a space-time integral over a function V of local fields) then

$$d\mu_V = d\mu_0 \exp\left(-\int V\right)$$

again satisfies Eq. (1.14) as

$$\sum_{i,k} c_i c_k^* Z_V[J_i - J_k] = \int d\mu_0 \exp\left(-\int V\right) \left| \sum_k c_k \exp(i(J_k, X)) \right|^2 \geq 0$$

with

$$Z_V[J] = \int d\mu_0(X) \exp\left(-\int V\right) \exp(i(J, X)).$$

Moreover, any limit of $d\mu_V$ will satisfy the inequality (1.14) allowing to claim that the limit defines a measure. This is a method to construct quantum field theories as perturbations of a Gaussian measure .

1.4   Groups and Their Representations

The group is a manifold G with a multiplication $\circ$ defined as a map $\circ : G \times G \to G$ satisfying the properties

(i) $g_1 \circ (g_2 \circ g_3) = (g_1 \circ g_2) \circ g_3$
(ii) there is a unit elément $e \in G$ such that for each $g \in G$ we have $g \circ e = e \circ g = g$
(iii) for each $g \in G$ there exists an element $g^{-1} \in G$ such that $g^{-1} \circ g = g \circ g^{-1} = e$.

We shall restrict ourselves mainly to groups which are matrices. In general we say that a group G has a representation U as a linear operator in a Hilbert space if there is a mapping $g \to U(g)$ such that

$$U(g_1 \circ g_2) = U(g_1)U(g_2). \tag{1.18}$$

In particular, if the Hilbert space has a finite dimension then U is a matrix.

A one parameter subgroup of the group is a map $R \to G$ such that

$$g(t_1) \circ g(t_2) = g(t_1 + t_2).$$

For the one parameter subgroup we can define its generator as

$$A = \frac{d}{dt} g(t)_{|t=0}.$$ (1.19)

It can be shown that if the manifold G has dimension n then there are n one parameter subgroups which define linearly independent operators A_j (vector fields, matrices for matrix groups) such that

$$[A_j, A_k] = f_{jkl} A_l.$$ (1.20)

The constants f characterize the group and are called the structure constants of the group [28].

As typical examples we might consider the group $O(n)$ of $n \times n$ orthogonal matrices (T denotes the transposition)

$$O^T O = 1$$

the group $U(n)$ of unitary matrices (+ is the hermitian conjugation)

$$U^+ U = 1$$

and the group $O(1, n-1)$ of matrices preserving the (Minkowski) scalar product

$$xy = x_0 y_0 - \mathbf{xy} \equiv \eta^{\mu\nu} x_\mu y_\nu$$

where $\mathbf{x} \in R^{n-1}$. The matrix $O \in O(1, n-1)$ satisfies the constraint

$$O^T \eta O = \eta.$$

We can expand O around the unit matrix $O \simeq 1 + \omega^{\mu\nu} m_{\mu\nu}$ where the parameters satisfy $\omega^{\mu\nu} = -\omega^{\nu\mu}$. Then, $m_{\mu\nu}$ satisfy the algebra

$$[m_{\mu\nu}, m_{\sigma\rho}] = \eta_{\mu\sigma} m_{\nu\rho} - \eta_{\mu\rho} m_{\nu\sigma} + \eta_{\nu\rho} m_{\mu\sigma} - \eta_{\nu\sigma} m_{\mu\rho}.$$ (1.21)

1.5 Exercises

1.1 Prove

$$\frac{d\theta}{dt} = \delta(t).$$

1.2 For $p^2 = p_0^2 - \mathbf{p}^2$ and $\omega = \sqrt{\mathbf{p}^2 + m^2}$ show that

$$\delta(p^2 - m^2) = (2\omega)^{-1} (\delta(p_0 + \omega) + \delta(p_0 - \omega))$$

Hint: Prove that

$$\delta(f(x)) = \sum_j \delta(x - x_j)|f'(x_j)|^{-1},$$

where x_j are zeros of the function $f(x)$. Then, use $f(p_0) = p_0^2 - \omega^2$.

1.3 Show

$$\frac{\delta G(F(\phi))}{\delta\phi(x)} = \frac{dG}{dF}\frac{\delta F}{\delta\phi(x)}.$$

1.4 Show that

$$\frac{\delta\phi(x)}{\delta\phi(y)} = \delta(x - y).$$

1.5 Using Exercises 1.3 and 1.4 calculate $\frac{\delta G(\phi)}{\delta\phi(x)}$ and higher order derivatives of

$$G(\phi) = \int dx_1....dx_n b(x_1, \ldots, x_n)\phi(x_1)....\phi(x_n) \exp\left(\int dxdy\phi(x)\phi(y)B(x, y)\right)$$

1.6 Prove the commutation relations (1.21).

Chapter 2
Quantum Theory of the Scalar Free Field

Abstract In this chapter we discuss a quantization of the classical relativistic scalar field. The scalar field as a relativistic wave function was proposed initially by Schrödinger, Klein and Gordon. It was recognized shortly that it does not describe electrons in the hydrogen atom. Moreover, for a physical interpretation it needs quantization. The scalar field theory is the simplest model for quantization as it contains no redundant degrees of freedom. We treat it as Hamiltonian quantum mechanics with an infinite number of degrees of freedom. We refer to Wigner's theory of relativistic particles identifying the free scalar field theory as a realization of the unitary representation of the Poincare group describing free particles with a mass and no spin. The quantization is realized in an abstract Fock space. In the final section we follow the Schrödinger prescription for quantization in the space of functions defined on the configuration space of fields. In this case the space of field configurations has an infinite number of dimensions.

In this chapter we discuss a quantization of the classical relativistic scalar field. The scalar field as a relativistic wave function was proposed initially by Schrödinger Klein and Gordon. It was recognized shortly that it does not describe electrons in the hydrogen atom. Moreover, it admitted negative particle energies (holes) and transitions between negative and positive energies. An interpretation of these transitions required field quantization. Then, soon after the discovery of a quantum description of the atom the need of a quantization of the electromagnetic field appeared. The discovery of the quantum nature of the electromagnetic field in the description of the black body radiation and in the photoelectric effect signified that the electromagnetic field in the field-atom interaction must also be treated as a quantum operator (see Dirac's paper [67]). In the paper of Born et al. [39] the polarization of the electromagnetic field was ignored. In such a description the quantization of the electromagnetic field was treated as a problem of the quantization of vibrations of a scalar field of a string. The scalar field appeared later in Yukawa theory of nuclear forces [248] which were supposed to be transmitted by scalar particles (pions). The scalar field theory is the simplest model for quantization as it contains no redundant degrees of freedom. We treat it as Hamiltonian mechanics with an infinite number of degrees of freedom. We can apply the same rules of quantization as for a non-relativistic particle

Z. Haba, *Lectures on Quantum Field Theory and Functional Integration*,
https://doi.org/10.1007/978-3-031-30712-6_2

described in the Hamiltonian formalism by a Hamiltonian with the canonical Pois-son brackets replaced by commutators (Sect. 2.2). The infinite number of degrees of freedom allows for an appearance of symmetries (Sect. 2.1) which are not present in mechanics. Then, in Sect. 2.3 we show that the quantized theory describes free particles. We refer to Wigner's theory of relativistic particles identifying the scalar field theory as a realization of the representation of the Poincare group describing free particles with no spin and defined mass. In Sect. 2.3 the quantization is realized in an abstract Fock space. In the final section we follow the Schrödinger prescription for quantization in the space of functions defined in the configuration space of fields. In this case the space of field configurations has an is infinite number of dimensions. An analysis in infinite dimensional spaces is necessary to work in the Schrödinger representation.

2.1 Classical Field Theory. Lagrange Equations and The Noether Theorem

The Lagrangian $\mathcal{L}$ is assumed to depend on fields ϕ (defined on the four dimensional Minkowski space M) and its derivatives. A necessary condition for the action

$$W = \int d^4x \mathcal{L} \tag{2.1}$$

to achieve an extremum (see [189] for a precise formulation) is a fulfillment of the Lagrange equation

$$\partial_\mu \frac{\partial \mathcal{L}}{\partial \partial_\mu \phi} = \frac{\partial \mathcal{L}}{\partial \phi}. \tag{2.2}$$

Let us now consider an infinitesimal transformation of the fields (we assume that ϕ is a vector with multiple components $\phi = (\phi^1, .., \phi^n)$, a multiplication of multi-component objects is understood as a multiplication and subsequent summation of components)

$$\phi' = \phi + \delta\phi. \tag{2.3}$$

Assume that $\mathcal{L}(\phi') = \mathcal{L}(\phi)$ then

$$\delta\mathcal{L} = \frac{\partial \mathcal{L}}{\partial \phi}\delta\phi + \frac{\partial \mathcal{L}}{\partial \partial_\mu \phi}\partial_\mu \delta\phi = \partial_\mu \left(\frac{\partial \mathcal{L}}{\partial \partial_\mu \phi}\delta\phi \right) = 0, \tag{2.4}$$

where in the second step we have applied the Lagrange equations (2.2). It follows that the current

$$J^\mu = \frac{\partial \mathcal{L}}{\partial \partial_\mu \phi}\delta\phi \tag{2.5}$$

is conserved, i.e., $\partial_\mu J^\mu = 0$. As a consequence the charge Q

$$Q = \int d\mathbf{x} J^0 \tag{2.6}$$

is time-independent (a constant).

Next, let us consider space-time symmetries. The simplest is the translation $x \rightarrow x + \epsilon$. The field is changing as

$$\phi' = \phi + \partial_\mu \phi \epsilon^\mu. \tag{2.7}$$

The action $W = \int d^4 x \mathcal{L}$ does not change but the Lagrangian is changing as

$$\delta\mathcal{L} = \epsilon^\mu \partial_\mu \mathcal{L}. \tag{2.8}$$

Then, on the left hand side of Eq. (2.4) we have $\delta\mathcal{L}$ of Eq. (2.8). As a consequence the conserved current is

$$T^\mu_\nu = \frac{\partial \mathcal{L}}{\partial \partial_\mu \phi} \partial_\nu \phi - \delta^\mu_\nu \mathcal{L}. \tag{2.9}$$

The conserved charge is the four-momentum

$$P_\mu = \int d\mathbf{x} T^0_\mu. \tag{2.10}$$

Under more general transformations (2.3) assume that the action $W = \int d^4 x \mathcal{L}$ does not change but the Lagrangian is changing as

$$\delta\mathcal{L} = \partial_\mu S^\mu. \tag{2.11}$$

Then, on the left hand side of Eq. (2.4) we have $\partial_\mu S^\mu$. Comparing both sides of Eq. (2.4) we conclude that the conserved current is

$$J^\mu = \frac{\partial \mathcal{L}}{\partial \partial_\mu \phi} \delta\phi - S^\mu, \tag{2.12}$$

where $\delta\phi$ as well as S depend on the parameters of the space-time transformations giving as many currents as there are parameters. As an example consider the infinitesimal Lorentz rotation (Sect. 1.4, Eq. (1.21)) $x^\mu \rightarrow x^\mu + \omega^{\mu\nu} x_\nu$ then

$$\delta\phi = \omega^{\mu\nu} x_\nu \partial_\mu \phi. \tag{2.13}$$

As conserved charges we obtain the angular momenta . After quantization the charges become the generators of the Lorentz transformations which satisfy the O(1, 3) algebra (1.21). The charges act upon the quantum fields leading to unitary representations

of the Lorentz group and translations (the Poincare group) in the Hilbert space of
quantum fields.

As an example of an internal symmetry we may consider the Lagrangian built
of two fields $\phi = (\phi_1, \phi_2)$ which is invariant under the transformation (the $O(2)$
rotation)

$$\phi_1' = \cos\alpha\phi_1 + \sin\alpha\phi_2,$$

$$\phi_2' = \cos\alpha\phi_2 - \sin\alpha\phi_1. \tag{2.14}$$

Then, assuming the invariance of the Lagrangian we obtain from Eq. (2.5) that the
current is

$$J^\mu = \frac{\partial\mathcal{L}}{\partial\partial_\mu\phi_1}\phi_2 - \frac{\partial\mathcal{L}}{\partial\partial_\mu\phi_2}\phi_1. \tag{2.15}$$

The charge resulting from the current (2.15) can be associated with the electric
charge.

2.2   Classical Scalar Free Field

The Lagrangian for the free scalar field of mass m is (we put in this section the
velocity of light $c = 1$)

$$\mathcal{L} = \frac{1}{2}\eta^{\mu\nu}\partial_\mu\phi\partial_\nu\phi - \frac{m^2}{2}\phi^2, \tag{2.16}$$

where $(\eta) = (1, -1, -1, -1)$. The Lagrange equations are (Klein-Gordon equations)

$$\eta^{\mu\nu}\partial_\mu\partial_\nu\phi + m^2\phi = 0. \tag{2.17}$$

The general solution of Eq. (2.17) is a superposition of plane waves (f_p and f_p^*)

$$f_p(x) = (2\pi)^{-\frac{3}{2}}\exp(-ipx), \tag{2.18}$$

where $px = p_0 x_0 - \mathbf{p}\mathbf{x}$ and $p_0 \equiv \omega(\mathbf{p}) = \sqrt{\mathbf{p}^2 + m^2}$.

We introduce the scalar product

$$(f, g) = i\int d\mathbf{x}(f^*\partial_0 g - g\partial_0 f^*). \tag{2.19}$$

It can be checked that if f and g satisfy the Klein-Gordon equation (2.17) then the
scalar product does not depend on time.

For the plane waves (2.18) we obtain

$$(f_p, f_q) = 2\omega(\mathbf{p})\delta(\mathbf{p} - \mathbf{q}), \tag{2.20}$$

where the formula (see Sect. 1.1)

$$\int d\mathbf{x}\exp(-i(\mathbf{p}-\mathbf{q})\mathbf{x}) = (2\pi)^3\delta(\mathbf{p}-\mathbf{q}) \qquad (2.21)$$

has been applied.

The general solution of the Klein-Gordon equation (2.17) can be expressed as a superposition of plane waves

$$\phi(x) = \int \frac{d\mathbf{p}}{2\omega(\mathbf{p})}\Big(a(\mathbf{p})f_p(x) + a^*(\mathbf{p})f_p^*(x)\Big). \qquad (2.22)$$

The complex conjugation $*$ comes from the requirement that $\phi(x)$ is a real function. There is a reason to write the integration measure in the above form as $\frac{d\mathbf{p}}{2\omega(\mathbf{p})}$ is invariant with respect to the Lorentz transformations (Exercise 1.2 in Chap. 1).

From the Lagrangian (2.16) we can define the Hamiltonian H in the same way as in the classical mechanics

$$H(\Pi, \phi) = \int d\mathbf{x}\Pi\partial_0\phi - \int d\mathbf{x}\mathcal{L}, \qquad (2.23)$$

where the canonical momentum Π is defined as

$$\Pi = \frac{\partial\mathcal{L}}{\partial\partial_0\phi}.$$

We can also use the Noether theorem to define the Hamiltonian from the energy-momentum $T^{\mu\nu}$ as

$$H = \int d\mathbf{x}T^{00}. \qquad (2.24)$$

Using the Lagrangian (2.16) we calculate either from the canonical Hamiltonian (2.23) or from the energy-momentum (2.24)

$$H = \frac{1}{2}\int d\mathbf{x}\Big((\partial_0\phi)^2 + (\nabla\phi)^2 + m^2\phi^2\Big) = \frac{1}{2}\int d\mathbf{x}\Big(\Pi^2 + (\nabla\phi)^2 + m^2\phi^2\Big), \qquad (2.25)$$

where the canonical momentum is $\Pi = \partial_0\phi$. The expression (2.25) is similar to the Hamiltonian of the harmonic oscillator $H = \frac{1}{2}(\partial_t q)^2 + \frac{1}{2}\omega^2 q^2 = \frac{1}{2}(p^2 + \omega^2 q^2)$ where p is the particle's momentum. This similarity will be more explicit if we use the expansion in plane waves (2.22) in order to express H as (Exercise 2.2)

$$H = \int \frac{d\mathbf{q}}{2\omega(\mathbf{q})}\Big(\omega(\mathbf{q})a^*(\mathbf{q})a(\mathbf{q})\Big). \qquad (2.26)$$

For a set of n harmonic oscillators with frequencies ω_j we would have a sum of n Hamiltonians in Eq. (2.26) instead of an integral over $\mathbf{q}$. We could say that classical

field theory is nothing more than a classical mechanics with an infinite number of
degrees of freedom. However, there is something new in classical field theory. These
are the symmetries. So, there is the invariance of the Lagrangian with respect to
spatial translations which according to the Noether theorem of Sect. 2.1 leads to the
momentum as the constant of motion

$$P_k = \int d\mathbf{x} T_{0k} = \int d\mathbf{x} \partial_0 \phi \partial_k \phi. \qquad (2.27)$$

In terms of Fourier components (2.22)

$$P_k = \int \frac{d\mathbf{q}}{2\omega(\mathbf{q})} \Big(q_k a^*(\mathbf{q}) a(\mathbf{q}) \Big). \qquad (2.28)$$

There is still the invariance under the Lorentz transformations which owing to the
Noether theorem leads to the constants of motion (Exercise 2.1)

$$M_{\mu\nu} = \int d\mathbf{x} \Pi \Big((x_\mu \partial_\nu - x_\nu \partial_\mu) \phi \Big). \qquad (2.29)$$

2.3   Quantization of the Scalar Field

As the classical field theory is just the classical mechanics with an infinite number
of degrees of freedom we do not need to invent new rules for quantization. We just
apply the rules of quantum mechanics to each degree of freedom. In terms of the
coordinate ϕ and the canonical momentum Π this rule applied to the commutator is
($\phi(\mathbf{x})$ commutes with $\phi(\mathbf{y})$ and $\Pi(\mathbf{x})$ with $\Pi(\mathbf{y})$)

$$[\phi(\mathbf{x}), \Pi(\mathbf{y})] = i\hbar \delta(\mathbf{x} - \mathbf{y}) \qquad (2.30)$$

(we set $\hbar = 1$ later on). We could also use the creation and annihilation operators
applied in the description of the harmonic oscillator $a = \frac{1}{\sqrt{2\omega}}(p - i\omega q)$ in order to
impose the quantization rule for the Fourier components of the scalar field

$$[a(\mathbf{p}), a^+(\mathbf{q})] = 2\omega(\mathbf{p})\hbar\delta(\mathbf{p} - \mathbf{q}). \qquad (2.31)$$

Here, a^+ means the Hermitian adjoint to the operator a. The operators $a(\mathbf{q})$ and $a(\mathbf{p})$
commute (as well as $a^+(\mathbf{q})$ and $a^+(\mathbf{p})$). The quantization rules (2.30) and (2.31) are
equivalent. This can be checked using the expansion (2.22) in plane waves (Exercise
2.3).

As in quantum mechanics we must find operators satisfying the commutation
relations (2.30) and (2.31) in Hilbert space. In quantum mechanics there is a theorem
(Stone-von Neumann) that all realizations of such commutation relations are unitarily

equivalent. This is not true in the case of an infinite number of degrees of freedom. We find a realization for free fields. We could realize the momentum Π as a functional derivative with the Hilbert space as a space of functions of an infinite number of variables. This is possible but involves functional integration (we do it in Sect. 2.5). In this section we choose another realization of the Hilbert space: the Fock space. We assume that there exists a vector $|0>$ (the vacuum) such that

$$a(\mathbf{p})|0>= 0. \tag{2.32}$$

This assumption determines the Hilbert space constructed as a completion of the linear combination of vectors of the form (where f_n are square integrable functions)

$$|\psi >= \sum_n c_n \int d\mathbf{p}_1...d\mathbf{p}_n f_n(\mathbf{p}_1, ..., \mathbf{p}_n)|\mathbf{p}_1...\mathbf{p}_n >, \tag{2.33}$$

where (the n-particle state, symmetric under the exchange $p_j \to p_k$)

$$|\mathbf{p}_1...\mathbf{p}_n >= a^+(\mathbf{p}_1)...a^+(\mathbf{p}_n)|0 > . \tag{2.34}$$

The scalar product of two vectors $< \psi'|\psi >$ is determined by the commutation relation of the operators a (2.31) and the vacuum condition (2.32) . This can most easily be derived for one particle states

$$
\begin{aligned}
< \mathbf{p}|\mathbf{q} >&=< 0|a(\mathbf{p})a^+(\mathbf{q})|0 >=< 0|a(\mathbf{p})a^+(\mathbf{q}) - a^+(\mathbf{q})a(\mathbf{p})|0 > \\
&=< 0|[a(\mathbf{p}), a^+(\mathbf{q})]|0 >= 2\hbar\omega(\mathbf{p})\delta(\mathbf{p} - \mathbf{q}) < 0|0 > .
\end{aligned}
\tag{2.35}
$$

If we define

$$|f >= \int \frac{d\mathbf{p}}{2\omega(\mathbf{p})} f(\mathbf{p})|\mathbf{p} > \tag{2.36}$$

then (we skip $\hbar$ from now on)

$$< f'|f >= \hbar \int \frac{d\mathbf{p}}{2\omega(\mathbf{p})} f'^*(\mathbf{p}) f(\mathbf{p}). \tag{2.37}$$

We can calculate the vacuum correlation functions

$$\Delta^{(+)}(x - y) =< 0|\phi(x)\phi(y)|0 >= (2\pi)^{-3} \int \frac{d\mathbf{p}}{2\omega(\mathbf{p})} \exp(-ip(x - y)), \tag{2.38}$$

where $p(x - y) = p_0(x_0 - y_0)) - \mathbf{p}(\mathbf{x} - \mathbf{y})$ with $p_0 = \omega(\mathbf{p}) = \sqrt{\mathbf{p}^2 + m^2}$. Equation (2.38) can also be expressed in an explicitly Lorentz invariant way (see Exercise 1.2)

$$\Delta^{(+)}(x - y) = (2\pi)^{-3} \int d^4 p\theta(p_0)\delta(p^2 - m^2) \exp(-ip(x - y)), \tag{2.39}$$

where $\theta(t) = 1$ for $t \geq 0$ and $\theta = 0$ for $t < 0$, $p^2 = p_0^2 - \mathbf{p}^2$. We introduce a time-ordered product of fields which will be important later on

$$T(\phi(x)\phi(y)) = \theta(x_0 - y_0)\phi(x)\phi(y) + \theta(y_0 - x_0)\phi(y)\phi(x). \qquad (2.40)$$

We calculate its vacuum expectation values

$$< 0|T(\phi(x)\phi(y))|0 > = \theta(x_0 - y_0)\Delta^{(+)}(x - y) + \theta(y_0 - x_0)\Delta^{(+)}(y - x)$$

$$= (2\pi)^{-3} \int \frac{d\mathbf{p}}{2\omega(\mathbf{p})} \exp(-i\omega(\mathbf{p})|x_0 - y_0| + i\mathbf{p}(\mathbf{x} - \mathbf{y})) \equiv i\Delta_F(x - y).$$
$$(2.41)$$

Let us still calculate the commutator of fields

$$[\phi(x), \phi(y)] = \Delta^{(+)}(x - y) - \Delta^{(+)}(y - x) \equiv \Delta(x - y). \qquad (2.42)$$

It is important to note that the free scalar field satisfies the principle of Einstein causality: the fields which cannot be connected by the light signal, i.e., $(x - y)^2 < 0$, do not disturb each other, i.e. $[\phi(x), \phi(y)] = 0$ (Exercise 2.4; use the Lorentz invariance of the commutator to go to the frame where $x_0 = y_0$, in this frame it is easy to see that the commutator is zero).

We would like to calculate the higher order correlation functions. The method of calculation is in principle the same: we move all annihilation operators to the right so that when acting on the vacuum they give 0, what remains are the numbers from the commutators. The result is the following (only even correlation functions are non-zero)

$$< 0|\phi(x_1).....\phi(x_{2n})|0 >$$
$$(2.43)$$
$$= \sum_P < 0|\phi(x_{i_1})\phi(x_{i_2})|0 > ... < 0|\phi(x_{i_{2n-1}}).....\phi(x_{i_{2n}})|0 >,$$

where the sum is over all possible pairings P of indices.

Equation (2.43) can be proved by induction for general n directly using creation and annihilation operators. A simpler way is to calculate

$$< 0| \exp(\phi(f))|0 > = \exp\left(\frac{1}{2} \int dxdy f(x)f(y)\Delta^{(+)}(x - y)\right), \qquad (2.44)$$

where $\phi(f) = \int dx f(x)\phi(x)$. Then, we can calculate the correlation functions of any number of fields by functional differentiation over f of Eq. (2.44). In order to prove (2.44) we need an identity for an exponential of a sum of creation and annihilation operators (Exercise 2.6).

In a similar way we can calculate the time-ordered correlation functions. In the result just replace the two-point functions in Eq. (2.43) by time ordered two-point functions

$$< 0|T(\exp(\phi(f)))|0 >= \exp\left(\frac{i}{2}\int dx dy f(x)f(y)\Delta_F(x-y)\right). \quad (2.45)$$

2.4 The Poincare Group and Its Representations

In order to complete the quantum theory of the free field we need to prove that it is invariant under Lorentz transformations and translations (Poincare group). This means that there should exist a unitary representation of the Poincare group such that the fields transform covariantly

$$U(\Lambda)\phi(x)U^{-1}(\Lambda) = \phi(\Lambda^{-1}x), \quad (2.46)$$

$$U(a)\phi(x)U^{-1}(a) = \phi(x-a), \quad (2.47)$$

where $a \in R^4$ and Λ is the Lorentz transformation $\Lambda \in O(1,3)$, see Sect. 1.4.

If $M_{\mu\nu}$ are the generators of the Lorentz group and P_μ are the generators of translations then by differentiation of Eqs. (2.46)–(2.47) over group parameters we can derive the commutation relations

$$[M_{\mu\nu}, \phi(x)] = -i(x_\mu\partial_\nu - x_\nu\partial_\mu)\phi(x), \quad (2.48)$$

$$[P_\mu, \phi(x)] = -i\partial_\mu\phi(x). \quad (2.49)$$

We need to construct these generators as Hermitian operators in the Fock space. Then, by exponentiation we obtain the representation of the Poincare group. From Eq. (2.28) we have

$$P_\mu = \int \frac{d\mathbf{q}}{2\omega(\mathbf{q})}\left(q_\mu a^+(\mathbf{q})a(\mathbf{q})\right). \quad (2.50)$$

For rotations from Eq. (2.29)

$$M_{jk} = \int \frac{d\mathbf{q}}{2\omega(\mathbf{q})}\left(a^+(\mathbf{q})(q_j\partial_k - q_k\partial_j)a(\mathbf{q})\right) \quad (2.51)$$

and

$$M_{0j} = \int \frac{d\mathbf{q}}{2\omega(\mathbf{q})}\left(a^+(\mathbf{q})\omega(\mathbf{q})\partial_j a(\mathbf{q})\right).$$

There remains to explain what does this free field theory describe. For this purpose let us calculate

$$P_\mu|\mathbf{p}(1),\ldots,\mathbf{p}(n) >= \left(\sum_j p_\mu(j)\right)|\mathbf{p}(1),\ldots,\mathbf{p}(n) > \quad (2.52)$$

Equation (2.52) shows that the energy and momentum of the states $|\mathbf{p}(1), \ldots, \mathbf{p}(n) >$ are just sums of the energies and momenta corresponding to a product of single states $|\mathbf{p}_j >$. It follows that the states $|\mathbf{p}(1), \ldots, \mathbf{p}(n) >$ describe n-non-interacting relativistic particles obeying Bose statistics (the states are symmetric under interchange of momenta).

From the Poincare algebra we can built two invariant operators commuting with all generators of the Poincare algebra and characterizing representations of the group [243]. These operators are: the mass

$$P^\mu P_\mu$$

and the operator

$$W_\mu W^\mu$$

where (the Pauli-Lubanski vector)

$$W_\mu = \frac{1}{2}\epsilon_{\mu\nu\sigma\rho}P^\nu M^{\sigma\rho}$$

$W_\mu W^\mu$ in the frame $\mathbf{P} = 0$ describes an internal momentum of the particle (the spin). We have (from Eqs. (2.48)–(2.51)) that

$$P_\mu P^\mu |\mathbf{p} >= m^2 |\mathbf{p} > \qquad (2.53)$$

and

$$W_\mu W^\mu |\mathbf{p} >= 0. \qquad (2.54)$$

It follows that the creation operators create n-particle states of non-interacting particles of mass m and spin 0.

2.5 Functional Representation of Quantum Fields

Instead of the realization of the quantum field theory in the Fock space we could follow the procedure known from elementary quantum mechanics. We choose as the Hilbert space the space of functions of the fields ϕ. If we formally follow the approach of quantum mechanics then the scalar product would be

$$\int d\phi \Psi_1^*(\phi)\Psi_2(\phi).$$

Then, the canonical momentum is defined as

$$\Pi(\mathbf{x}) = -i\hbar \frac{\delta}{\delta\phi(\mathbf{x})}. \tag{2.55}$$

With this scalar product the annihilation operator is

$$a(\mathbf{k}) = \frac{1}{\sqrt{2\omega}}(-i\omega\phi(\mathbf{k}) + \Pi(\mathbf{k})),$$

where

$$\phi(\mathbf{k}) = (2\pi)^{-\frac{3}{2}} \int d\mathbf{x} \exp(-i\mathbf{k}\mathbf{x})\phi(\mathbf{x}).$$

In the functional representation the Hamiltonian has the form (up to an infinite renormalization constant)

$$H = \frac{1}{2} \int d\mathbf{x} \left(-\hbar^2 \frac{\delta^2}{\delta\phi(\mathbf{x})\delta\phi(\mathbf{x})} + (\nabla\phi)^2 + m^2\phi^2 \right). \tag{2.56}$$

However, the standard Lebesgue measure applied above for the definition of the scalar product has no extension to infinite dimensions. There is some arbitrariness in the choice of the Hilbert space in quantum mechanics (instead of the space $L^2(d\mathbf{x})$ we can use $L^2(\rho^2 d\mathbf{x})$ with a certain multiplier ρ^2). We define the scalar product by means of the Gaussian measure (see Sect. 1.3) which is well-defined in an infinite number of dimensions

$$(\Psi_1, \Psi_2) = \int d\mu_0(\phi)\Psi_1^*(\phi)\Psi_2(\phi), \tag{2.57}$$

where $d\mu_0(\phi)$ has the heuristic expression (its precise mathematical meaning is discussed in Sect. 1.3)

$$d\mu_0(\phi) = Z^{-2}d\phi \exp\left(-\frac{1}{\hbar}(\phi, \omega\phi) \right) \equiv d\phi|\psi_g|^2 \tag{2.58}$$

with $\omega = \sqrt{-\Delta + m^2}$. Z is a normalization factor such that $\int d\mu_0 = 1$. Here, ψ_g is the ground state of the (formal) Hamiltonian (2.56). In the above mentioned passage $L^2(d\mathbf{x}) \to L^2(\rho^2 d\mathbf{x})$ we change states as $\psi \to \rho^{-1}\psi$ and operators A

$$A \to \rho^{-1}A\rho. \tag{2.59}$$

After the transformation (2.59) (with $\rho = \psi_g$) the ground state is 1 and the Hamiltonian

$$H = \frac{1}{2} \int d\mathbf{x} \left(-\hbar^2 \frac{\delta^2}{\delta\phi(\mathbf{x})\delta\phi(\mathbf{x})} + 2\hbar(\omega\phi)(\mathbf{x}) \frac{\delta}{\delta\phi(\mathbf{x})} \right). \tag{2.60}$$

The canonical momentum instead of (2.55) takes the form

$$\Pi(\mathbf{x}) = -i\hbar\frac{\delta}{\delta\phi(\mathbf{x})} + i(\omega\phi)(\mathbf{x}). \tag{2.61}$$

H (2.60) and Π (2.61) are symmetric operators in $L^2(d\mu_0)$.

2.6 Exercises

2.1 Calculate the current corresponding to the invariance with respect to Lorentz transformations.

Hint: the infinitesimal Lorentz transformation can be represented as $\delta\phi = \omega^{\mu\nu}(x_\mu\partial_\nu - x_\nu\partial_\mu)\phi$ where $\omega^{\mu\nu} = -\omega^{\nu\mu}$.

2.2 Prove the formula (2.26) for the Hamiltonian in terms of Fourier components.

2.3 Show that the canonical quantization rule (2.30) and the creation-annihilation quantization (2.31) are equivalent.

2.4 Prove that $\Delta(x - y) = 0$ if $(x - y)^2 < 0$

Hint:use Lorentz invariance and the fact that space-like separated events can be made simultaneous in a certain frame.

2.5 Prove Eqs. (2.53)–(2.54).

2.6 Show that if

$$[A, [A, B]] = [B, [A, B]] = 0$$

then

$$\exp(A + B) = \exp(A)\exp(B)\exp\left(-\frac{1}{2}[A, B]\right).$$

Hint: multiply the operators by a parameter t and derive a differential equation satisfied by both sides.

2.7 Using Exercise 2.6 calculate $< 0|\exp(\phi(f))|0 >$.

2.8 Show that $[W_\mu W^\mu, P_\nu P^\nu] = 0$ and that $W_\mu W^\mu$ and $P_\nu P^\nu$ commute with all generators of the Poincare algebra. Prove Eqs. (2.53)–(2.54). Hint: express the creation operator by ϕ and use Eqs. (2.46)–(2.47).

2.9 Derive Eq. (2.60) from (2.56) and (2.59) when $\rho = \psi^g$ is the ground state.

Chapter 3
Interacting Fields and Scattering Amplitudes

Abstract We quantize the interaction of scalar fields in the interaction picture. The benefit of working in the interaction picture consists in the simple way of the derivation of the Gell-Mann-Low formula for quantum field correlation functions. The calculation at the second order of perturbation theory reveals the divergencies which can be cured by the mass and coupling constant renormalization. We show that divergencies are to be expected in higher orders of scalar field theories in d-dimensions (power counting). Then, we discuss the Heisenberg picture in the rigorous mathematical formulation of Lehmann, Symanzik and Zimmermann which leads to an expression of the scattering amplitude by the time ordered vacuum correlation functions of quantum fields. Then, the Gell-Mann-Low formula of the interaction picture allows to calculate the scattering amplitude in perturbation theory (Feynman diagrams). In this chapter we follow mainly the standard old-fashioned approach to QFT, but we outline also a continuation to an imaginary time (Euclidean field theory) which allows an application of some modern methods developed in the seventies and eighties which are discussed in subsequent chapters.

The field quantization was primarily introduced to describe a particle interaction with radiation quanta. In the Heisenberg picture one can view this issue as the problem of finding quantum fields satisfying a set of non-linear equations. In the perturbative calculations it was recognized soon (Heisenberg and Pauli [150]) that the second order results lead to divergencies. The problem remained unsolved till the post war time (see Weinberg [240] for more detailed history) when the method of renormalization has been developed. We are going to quantize the scalar fields in the interaction picture although a direct method of solving operator partial differential equations (Heisenberg picture) is also possible (see Sect. 3.11). The benefit of working in the interaction picture results from the simple way of the derivation of the Gell-Mann-Low formula for quantum field correlation functions (Sect. 3.1). The calculation at the second order of perturbation theory reveals the divergencies which can be cured by the mass and coupling constant renormalization (Sect. 3.6). We show in Sect. 3.8 that divergencies are to be expected in higher orders of scalar field theories in d-dimensions (power counting). In the interaction picture we can calculate the scattering amplitude directly from the Hamiltonian (Sect. 3.1). However, the rigorous mathematical formulation of Lehman, Symanzik and Zimmerman leads to an expression of the scattering amplitude by the time-ordered vacuum correlation functions of quantum fields (Sect. 3.12). Then, the Gell-Mann-Low formula of the interaction picture allows to calculate the scattering amplitude in perturbation theory (Feynman diagrams). In a relativistic quantum field theory the interaction picture is false (Haag's theorem [116, 131]). However, the Gell-Man-Low formula is correct after an analytic continuation to imaginary time (Euclidean field theory). We discuss the Osterwalder-Schrader results on the relation between real and imagnary time in Sect. 3.10.

We follow in this chapter mainly the standard old-fashioned approach to QFT as presented in many text-books concerning QFT and its applications to particle physics [36, 53, 102, 196, 214, 240]. In subsequent chapters we relate this standard approach to some modern methods developed in the seventies and eighties which are not treated in the above mentioned text-books.

3.1   Interaction Picture

We can solve Eq. (2.49) (with an initial condition at $t = 0$)

$$\phi(t, \mathbf{x}) = \exp(i P_0 t) \phi(t = 0, \mathbf{x}) \exp(-i P_0 t). \tag{3.1}$$

If $P_0 = H_0 + V$ (with H_0 defined either by (2.50) or (2.60) and ϕ by Eq. (2.22) or as a multiplication operator in the functional representation of Sect. 2.5) then it can be checked by formal calculations (without any care whether the expressions are finite) that this ϕ (3.1) satisfies equations of motion

$$\partial^\mu \partial_\mu \phi + m^2 \phi + V'(\phi) = 0.$$

From the classical Lagrangian (and Hamiltonian) for the interaction $V(\phi)$ we have $H(\phi(t, \mathbf{x})) = H(\phi(t = 0, \mathbf{x}))$. Hence, we can insert in the Hamiltonian the initial fields. We have

$$P_0 = H = H_0 + \int d\mathbf{x} V(\phi(\mathbf{x})) \equiv H_0 + H_I(0), \tag{3.2}$$

where H_0 is the free Hamiltonian.

Let us denote the free field discussed in Chap. 2 by ϕ_{in} (incoming field). We have for the free field

$$\phi_{in}(t, \mathbf{x}) = \exp(i H_0 t) \phi(t = 0, \mathbf{x}) \exp(-i H_0 t). \tag{3.3}$$

Define the interaction picture evolution as

$$U_I(t) = \exp(i H_0 t) \exp(-i H t). \tag{3.4}$$

By direct differentiation we obtain

$$\partial_t U_I = -i \exp(i H_0 t)(H - H_0) \exp(-i H_0 t) U_I. \tag{3.5}$$

Let us define

$$H_I(t) = \exp(i H_0 t) V(\phi(t = 0, \mathbf{x})) \exp(-i H_0 t) = V(\phi_{in}(t, \mathbf{x})). \tag{3.6}$$

Then, the evolution equation (3.5) reads

$$\partial_t U_I = -i H_I(t) U_I. \tag{3.7}$$

We can solve this equation (with the initial condition $U(0) = 1$) by iteration. In the lowest order we get

$$U_I = 1 - i \int_0^t ds\, H_I(s).$$

Repeating the iteration we obtain

$$U_I = 1 - i \int_0^t ds\, H_I(s) + \cdots (-i)^n \int_0^t ds_n ... \int_0^{s_2} ds_1 H_I(s_n)....H_I(s_1) + \cdots$$

$$= 1 - i \int_0^t ds\, H_I(s) + \cdots (-i)^n (n!)^{-1} \int_0^t ds_n ... \int_0^t ds_1 T(H_I(s_n)....H_I(s_1)) + \cdots$$

$$\equiv T\left(\exp(-i \int_0^t H_I(s) ds) \right), \tag{3.8}$$

where the time ordering is a generalization of the one of Sect. 2.3 (Eq. (2.40)) saying that the operators with bigger time are moved to the left. We replace the integration in Eq. (3.8) at the lower limit from 0 to $-\infty$ (as we change the initial condition to $-\infty$). We could do it from the beginning, taking the initial condition in Eq. (3.1) at $t_0 = -\infty$ in

$$\phi_{in}(t, \mathbf{x}) = \exp(i H_0(t - t_0))\phi(t_0, \mathbf{x}) \exp(-i H_0(t - t_0)). \tag{3.9}$$

Then, the scattering matrix S is defined as the evolution from $-\infty$ to $+\infty$ (in this way the scattering operator is defined in elementary quantum mechanics)

$$S = T\left(\exp\left(-i \int_{-\infty}^{\infty} H_I(s)ds \right) \right). \tag{3.10}$$

With this definition the operator S is invariant under the transformations (2.46)–(2.47) of the Poincare group. We shall treat the expression for the time-ordered exponential (3.10) in terms of the perturbation series (3.8) although it also has a non-perturbative meaning (product integral) [70] as the limit of the product

$$\prod_j \exp(-i\,H_I(s_j)\Delta s_j)$$

when $\Delta s_j \to 0$. We can now take as the initial state the k particle state constructed by means of the creation operators of Chap. 2 and calculate the probability amplitude that the final state will consist of r particles

$$S_{rk} = < r|S|k > \tag{3.11}$$

(see Exercise 3.2; the scattering amplitude for $V = g\phi^4$).

We immediately encounter a difficulty in the calculation of the scattering amplitude: an infinity resulting from infinite vacuum expectation values, e.g., in ϕ^4

$$< 0|\phi^4(x)|0 >= 3(\Delta^{(+)}(0))^2 = \infty \tag{3.12}$$

because

$$\Delta^{(+)}(0) = (2\pi)^{-3} \int d\mathbf{p}(2\omega(\mathbf{p}))^{-1} = \infty. \tag{3.13}$$

This is a trivial infinity which we can remove by replacing ϕ^4 by "normal ordered powers" $: \phi^4 :$ (so in the Exercise 3.2 it is understood that we ignore the infinity (3.12), removing it by normal ordering). The normal ordering is defined in such a way that its vacuum expectation value is equal to zero. The definition of any power can be obtained from a definition of the normal ordered exponential

$$: \exp(\lambda\phi(x)) := \exp(\lambda\phi(x))(< 0| \exp(\lambda\phi(x))|0 >)^{-1}$$

$$\tag{3.14}$$

$$= \exp(\lambda\phi(x)) \exp(-\tfrac{\lambda^2}{2}\Delta^{(+)}(0)),$$

where we have calculated the vacuum expectation value on the rhs using the formula (2.44) of Sect. 2.3. The definition is by means of an infinite expression $\Delta^+(0)$ but we can make it finite applying it to fields with a cut-off on momenta (restricting the integration over momenta in Eq. (3.13) to a finite interval). The crucial point is that after a calculation of correlation functions of regularized normal powers we obtain finite expressions after the removal of the cut-off. From Eq. (3.14) we calculate normal powers by differentiation over λ and subsequently letting $\lambda = 0$. It follows directly from the definition (3.14) that the vacuum expectation value of the normal power is zero. The correlation functions of the normal powers (also called Wick powers) can be calculated using Eqs. (3.14) and (2.44)

$$< 0| : \phi^2 : (x) : \phi^2 : (y)|0 >= c_2(\Delta^{(+)}(x-y))^2, \tag{3.15}$$

$$< 0| : \phi^4 : (x) : \phi^4 : (y)|0 >= c_4(\Delta^{(+)}(x-y))^4. \tag{3.16}$$

In a similar way we can calculate the time ordered correlation ,e.g.,

$$< 0|T(: \phi^4 : (x) : \phi^4 : (y))| >= c_4(i\Delta_F(x-y))^4. \tag{3.17}$$

As an Exercise 3.1: prove the formulas (3.15)–(3.17)and calculate the coefficients c_n. In the calculations use the exponential definition (3.14), Eqs. (2.44)–(2.45) and a differentiation over λ in Eq. (3.14) in order to generate the normal ordered powers.

3.2 Correlation Functions

We can define the field in the Heisenberg picture as the solution

$$\phi(t, \mathbf{x}) = \exp(i P_0(t - t_0))\phi(t_0, \mathbf{x}) \exp(-i P_0(t - t_0)) \tag{3.18}$$

of the equation

$$i\partial_t \phi = [\phi, P_0]. \tag{3.19}$$

Later we shall take this initial t_0 as $t_0 = -\infty$ and call the initial field "incoming". From Eqs. (3.1) and (3.4) we obtain

$$\phi(t, \mathbf{x}) = U_I(t, t_0)^+ \phi_{in}(t, \mathbf{x})U_I(t, t_0). \tag{3.20}$$

Here, $U_I(t, t_0)$ is the solution of the evolution equation (3.7) with the initial condition at t_0. As every evolution operator

$$U_I(t, t_0) = U_I(t, t_1)U_I(t_1, t_0). \tag{3.21}$$

Let $|\Phi_0 >$ be the physical vacuum

$$H|\Phi_0 >= 0 \tag{3.22}$$

and $|0 >$ the Fock vacuum (for the free field)

$$H_0|0 >= 0. \tag{3.23}$$

Assume that $|\Phi_0 >$ as well as all physical states are constructed in the Fock space. We can express the physical vacuum by the Fock vacuum

$$|\Phi_0 >= A \lim_{\tau \to -\infty} \exp(iH\tau) \exp(-iH_0\tau)|0 >= A \lim_{\tau \to -\infty} U_I^+(\tau)|0 >, \qquad (3.24)$$

where we use the formula (3.9) for U_I. The argument is the following: the eigenstates of H form a basis of the Hilbert space, hence (we do not normalize the states Φ_n)

$$|0 >= \frac{1}{A}|\Phi_0 > + \sum_{n>0} c_n|\Phi_n >, \qquad (3.25)$$

where $|\Phi_n >$ are the energy eigenstates (with eigenvalues ϵ_n). So

$$\exp(iH\tau) \exp(-iH_0\tau)|0 >= \exp(iH\tau)|0 >$$
$$= \frac{1}{A}|\Phi_0 > + \sum_n c_n \exp(i\epsilon_n\tau)|\Phi_n > . \qquad (3.26)$$

When $\tau \to -\infty$ the oscillating terms in the mean tend to zero. A justification of the limit comes from the Lebesgues lemma which says that

$$lim_{\tau \to \infty} \int d\epsilon \exp(i\epsilon\tau) f(\epsilon) = 0 \qquad (3.27)$$

if $|f(\epsilon)|$ is an integrable function (the spectrum of multi-particle states in Eq. (3.26) is continuous).

In order to determine the constant A multiply Eq. (3.24) by $< \Phi_0|$. We obtain (assuming $< \Phi_0|\Phi_0 >= 1$)

$$A = \left(\lim_{\tau \to -\infty} < \Phi_0|U_I(\tau)^+|0 > \right)^{-1}. \qquad (3.28)$$

It can be seen that we can replace $\tau \to -\infty$ by $\tau \to \infty$ in Eqs. (3.24)–(3.27) so that we can write

$$A^* = \left(\lim_{\tau' \to \infty} < 0|U_I(\tau')|\Phi_0 > \right)^{-1}. \qquad (3.29)$$

Let us consider

$$U_I(\tau')U_I^+(\tau) = \exp(iH_0\tau') \exp(-iH\tau') \exp(iH\tau) \exp(-iH_0\tau). \qquad (3.30)$$

For large τ, τ' by means of the argument used at (3.26) we obtain

$$< 0|U_I(\tau')U_I^+(\tau)|0 >\simeq< 0|U_I(\tau')|\Phi_0 >< \Phi_0|U_I^+(\tau)|0 >$$
$$=< 0| \exp(-iH\tau')|\Phi_0 >< \Phi_0| \exp(iH\tau)|0 >\simeq (AA^*)^{-1}. \qquad (3.31)$$

Hence, if we let $\tau' \to \infty$ and $\tau \to -\infty$ then $U_I(\tau')U_I^+(\tau) \to U(\infty, -\infty)$ and

$$|A|^{-2} = < 0|U_I(\infty, -\infty)|0 > . \tag{3.32}$$

Remark: In Eq. (3.20) we expressed the interacting field by the free field. Using Eq. (3.24) we can calculate the correlation functions of the interacting fields in the true physical vacuum in terms of the free fields and their correlation functions in the Fock space vacuum. We do it in the next section. Strictly speaking according to the Haag theorem [116, 131] the interacting field and the free field cannot (in relativistic QFT) satisfy the canonical commutation relations in the same Hilbert space. Hence, the interaction picture (without the momentum and volume cutoffs violating the relativistic invariance) is false. The Euclidean version recovers (in an imaginary time) formulas derived by means of the interaction picture. The perturbation series for correlation functions derived in the interaction picture holds true in non-perturbative approaches to QFT as for example in $P(\phi)_2$ models [111].

3.3 Gell-Mann-Low Formula

We are interested in calculation of the generating functional (where $J(x)$ is a real function on R^4)

$$Z[J] = 1 + \sum_{n=1}^{\infty} \frac{i^n}{n!} \int dx_1...dx_n S_n(x_1, \ldots, x_n)J(x_1)...J(x_n), \tag{3.33}$$

where

$$S_n(x_1, \ldots, x_n) = < \Phi_0|T(\phi(x_1).......\phi(x_n))|\Phi_0 >, \tag{3.34}$$

and Φ_0 is the physical vacuum $H\Phi_0 = 0$. Let us consider first

$$W_n(x_1, ..., x_n) = < \Phi_0|\phi(x_1).......\phi(x_n)|\Phi_0 > . \tag{3.35}$$

Using Eq. (3.24) we express the physical vacuum Φ_0 by the Fock vacuum on the rhs of the vacuum expectation value (3.35) and on the lhs of this expectation value we apply Eq. (3.24) with $\tau' \to +\infty$. Then, the normalization constant is given by Eq. (3.32). We express the interacting fields by the free (incoming) fields ϕ_{in} using Eq. (3.20)

$$\begin{aligned}
&W_n(x_1, \ldots, x_n) \\
&= AA^* < 0|U_I(\infty)U_I^+(t_1)\phi_{in}(x_1)U_I(t_1)U_I^+(t_2)\phi_{in}(x_2)U_I(t_2) \\
&........\phi(x_n)U_I(t_n)U_I^+(-\infty)|0 > .
\end{aligned} \tag{3.36}$$

Here $U_I(\infty) = U_I(\infty, 0)$ and $U_I^+(-\infty) = U_I^+(-\infty, 0) = U_I(0, -\infty)$. Next, we consider the time-ordered product in Eq. (3.35) and we use the composition law (3.21)

$$U_I(t, t_0) = U_I(t, t_1)U_I(t_1, t_0).$$

Inside the time-ordered product we can exchange the order of operators. Then, the product of U_I is

$$U_I(\infty)U_I^+(t_1)U_I(t_1)U_I^+(t_2)U_I(t_2)\ldots\ldots U_I(t_n)U_I^+(-\infty) = U_I(\infty, -\infty).$$

Inserting the expression (3.32) for AA^* we obtain the final formula (Gell-Mann-Low [36, 102, 106])

$$\begin{aligned}
S_n(x_1, \ldots, x_n) &= (< 0|U_I(\infty, -\infty)|0 >)^{-1} \\
&< 0|T\Big(U_I(\infty, -\infty)\phi_{in}(x_1)\ldots\ldots\phi_{in}(x_n)\Big)|0 > .
\end{aligned} \tag{3.37}$$

For the generating functional (3.33) we have from Eq. (3.37)

$$\begin{aligned}
Z[J] &=< \Phi_0|T\Big(\exp(i \int dx J(x)\phi(x))\Big)|\Phi_0 > \\
&=< 0|T\Big(U_I(\infty, -\infty)\exp(i \int dx J(x)\phi_{in}(x))\Big)|0 > \Big(< 0|U_I(\infty, -\infty)|0 >\Big)^{-1}.
\end{aligned} \tag{3.38}$$

Note that the last factor in Eq. (3.38) is necessary for the correct normalization $Z[0] = 1$.

We know that U_I from Eq. (3.10) can be expressed by free fields (cp., Eq. (3.6)). Let us consider the ϕ^4 model (which is the only non-trivial renormalizable model of scalar fields in four dimensions with positive energy, see Sect. 3.8)

$$H_I(t) = g \int d\mathbf{x} : \phi_{in}(x)^4 :$$

and

$$U_I = T\Big(\exp\Big(-ig \int d^4x : \phi_{in}^4 :\Big)\Big).$$

In an expansion of U_I in power series of g we have the four-dimensional (Lorentz invariant) integrals $\int d^4x : \phi_{in}(x)^4 :$.

Let us calculate in the lowest order the four-point function (we skip the index "in")

$$\begin{aligned}
S_4^{(1)}(x_1, \ldots, x_4) &= -ig \int d^4y \\
&< 0|T(: \phi(y)^4 : \phi(x_1)\ldots\ldots\phi(x_4))|0 > .
\end{aligned}$$

The calculation goes according to the formula for free fields (2.43) of Sect. 2.3 (but pairings inside $: \phi^4 : (y)$ are absent owing to the normal ordering)

$$S_4^{(1)}(x_1,\ldots,x_4) = -ig \int d^4y \Delta_F(y-x_1)\Delta_F(y-x_2)\Delta_F(y-x_3)\Delta_F(y-x_4)$$
$$(3.39)$$

Thanks to the normal ordering of ϕ^4 we have no infinities which would come from $\Delta_F(y-y)$.

At the second order (see Eq. (3.17))

$$S_4^{(2)}(x_1,\ldots,x_4) = (-ig)^2 \int d^4y_1 d^4y_2$$
$$\Delta_F(y_1-x_1)\Delta_F(y_1-x_2)\Delta_F(y_2-x_3)\Delta_F(y_2-x_4) \qquad (3.40)$$
$$(i\Delta_F(y_1-y_2))^2 + \cdots,$$

where the omitted terms correspond to the remaining pairings between the points $y_1, y_2, x_1, \ldots, x_4$.

To an arbitrary $\mathcal{V}$-order for E-point functions the rule is the following (Feynman diagrams):

1. pick $\mathcal{V}$-points $y_1, \ldots, y_\mathcal{V}$ as vertices
2. pick E external points $x_1, \ldots, x_E$
3. connect all external points with vertices (without crossing) and vertices among themselves (in ϕ^N theory there are N lines entering each vertex).
4. multiply the propagators $i\Delta_F(x_j - y_k)$ associated to the line connecting x_j with y_k and $i\Delta_F(y_j - y_k)$ connecting the vertices.
5. multiply the whole expression by $(-ig)^\mathcal{V}$
6. integrate over $y_1, y_2, \ldots, y_\mathcal{V}$
7. sum up over all possible diagrams.
8. we do not take into account the disconnected diagrams corresponding to closed loops connecting the vertices y_j because such diagrams cancel with the contributions from the denominator of Eq. (3.37).

The Feynman rules corresponding to Eqs. (3.39) and (3.40) are depicted below

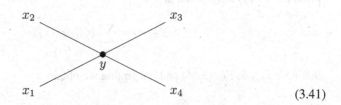

$$(3.41)$$

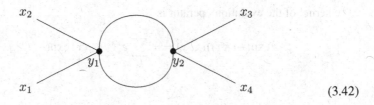

$$(3.42)$$

3.4   The Integral Kernel of an Operator

In subsequent sections we need more information about propagators as kernels, i.e., as integral operators corresponding to some differential operators. We define an integral operator in $L^2(dx)$ [185] by means of the formula

$$(K\psi)(x) = \int K(x, y)\psi(y)dy. \tag{3.43}$$

$K(x, y)$ is called the integral kernel of the operator K. Every linear operator can be expressed as an integral operator if we admit the kernels which are distributions [185]. Directly from the definition it follows

1. $1(x, y) = \delta(x - y)$
   (the kernel of the unit operator is the δ-function)
2. If $K = K_1 K_2$ then
   $K(x, y) = \int dz K_1(x, z) K_2(z, y)$
3. $(K^+)(x, y) = K^*(y, x)$

If we have orthonormalized eigenfunctions of an operator K, $K e_n = \lambda_n e_n$, (with $(e_n, e_m) = \delta_{nm}$) then

$$K(x, y) = \sum_n \lambda_n e_n^*(y) e_n(x). \tag{3.44}$$

In order to prove this formula it is sufficient to check that

$$(K e_m)(x) = \int dy \sum_n \lambda_n e_n^*(y) e_n(x) e_m(y) = \lambda_m e_m(x). \tag{3.45}$$

From Eq. (3.44) it follows that

$$\delta(x - y) = \sum_n e_n^*(y) e_n(x). \tag{3.46}$$

Setting $x = y$ in Eq. (3.44) and integrating we obtain

$$\int dx K(x, x) = \sum_n \lambda_n = Tr(K). \tag{3.47}$$

The kernel of the evolution operator is

$$(\exp(-iKt))(x, y) == \sum_n e_n^*(y) e_n(x) \exp(-it\lambda_n). \tag{3.48}$$

The formulas hold true in the sense of the convergence in the space of distributions which we do not explore here.

In order to determine the kernel K of an operator inverse to M we must solve the equation

$$MK = 1. \tag{3.49}$$

In terms of kernels

$$\int dz M(x, z) K(z, y) = \delta(x - y).$$

As an example: the kernel of $M = -\frac{d^2}{dx^2} + m^2$ is

$$\left(-\frac{d^2}{dx^2} + m^2 \right) \delta(x - z).$$

So Eq. (3.49) reads

$$\left(-\frac{d^2}{dx^2} + m^2 \right) K(x, y) = \delta(x - y).$$

If we have the Hilbert space $L^2(\rho^2 dx)$ (as for quantum mechanics on a manifold) then we could define the kernel in terms of the formula

$$(K\psi)(x) = \int \rho(x)^{-1} K(x, y) \rho(y)^{-1} \psi(y) \rho^2(y) dy. \tag{3.50}$$

With such a definition the kernel of an operator symmetric in $L^2(\rho^2 dx)$ has the kernel satisfying the condition $K(x, y) = K(y, x)^*$ as in $L^2(dx)$ and the kernel of the unit operator is still $\delta(x - y)$.

3.5 Momentum Representation

We have calculated $i\Delta_F$ in Chap. 2 in Eq. (2.41). Let us check by direct computations

$$-(m^2 + \partial_\mu \partial^\mu)\Delta_F(x) = \delta(x), \tag{3.51}$$

where $\delta(x)$ is the δ function in $S^*(R^4)$.

For the calculations all we need to know is

$$\frac{d}{dt}\theta(t) = \delta(t)$$

and

$$(2\pi)^{-3} \int d\mathbf{p} \exp(i\mathbf{p}(\mathbf{y} - \mathbf{x})) = \delta(\mathbf{y} - \mathbf{x}).$$

Let us take the Fourier transform of Eq. (3.51) and define

$$\Delta_F(x) = (2\pi)^{-4} \int dp \exp(-ipx) G_F(p). \tag{3.52}$$

Then, Eq. (3.51) takes the form

$$(p^2 - m^2) G_F = 1$$

with the solution $G_F = (p^2 - m^2)^{-1}$. Products and squares in Eqs. (3.51)–(3.52) are meant in the sense of the Minkowski scalar products and squares and by dp we denote the four-dimensional integral $d^4 p$.

We can calculate the propagator $\Delta_F(x)$ as the Fourier transform of $G_F = (p^2 - m^2)^{-1}$. The integral is not uniquely defined because of the singularity of the integral at $p_0 = \pm \omega(\mathbf{p})$. The integral can be performed by shifting the integration path from the real line to the complex plane and computing the integral by the method of residues taking only the pole at $p_0 = \omega(\mathbf{p})$ into account [214]. Then, the result agrees with the formula (2.41) (Exercise 3.3).

3.6 Coupling Constant Renormalization

If we take the Fourier transform of the four-point correlation function (Eq. (3.39)) at the first order of the coupling constant expansion in ϕ^4 theory then we obtain

$$- ig\delta(p_1 + p_2 - p_3 - p_4) G_F(p_1) G_F(p_2) G_F(p_3) G_F(p_4) \tag{3.53}$$

At the order g^2 we have from Eq. (3.40)

$$(-ig)^2 G_F(p_1) G_F(p_2) \int dp dq \delta(p_1 + p_2 - p - q) G_F(p) G_F(q)$$
$$\delta(p_3 + p_4 - p - q) G_F(p_3) G_F(p_4). \tag{3.54}$$

A generalization of this formula allows to formulate the Feynman rules for correlation functions in the momentum space (corresponding to the Feynman rules in coordinate space of Sect. 3.3) at the $\mathcal{V}$th order of the perturbative expansion with E external momenta for the E-point function:

(i) pick up $\mathcal{V}$ vertices and connect them by internal and external lines with (directed) momenta (in the ϕ^N model there are N lines entering or leaving each vertex)

(ii) associate to the external line with momentum p the propagator $G_F(p)$ and to the internal lines with momentum q the propagator $G_F(q)$

(iii) to each vertex associate $(-ig)\delta$- function of conserved momenta entering and leaving the vertex
(iv) integrate over internal momenta q
 (v) multiply by propagators of the external lines.

The rules corresponding to the results (3.53) and (3.54) (one-loop diagram) are depicted below (the diagrams were drawn by our PhD student Patryk Mieszkalski)

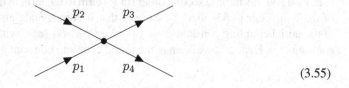

$$(3.55)$$

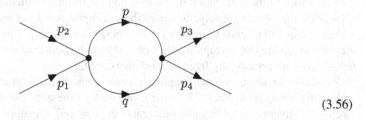

$$(3.56)$$

The rule that in $V(\phi) = g\phi^N$ model there are N lines entering or leaving the vertex follows from the Gaussian integration rule in Eq. (3.37) that we consider all pairings with the vertex represented by $: V(\phi(x)) :$. This is shown in the examples (3.39) and (3.40) for $g\phi^4$. As will be explained in Sect. 3.12 if we omit v) (multiplication by propagators of external lines) then the ("amputated") Feynman diagrams give us the rule for a calculation of the scattering amplitude. As in the coordinate space (3.40) we do not take into account the disconnected diagrams.

In Eq. (3.54) we have (after integrating δ-functions) integrals of the form ($k = p_1 + p_2$)

$$\int dq \, G_F(q) G_F(q-k). \qquad (3.57)$$

The q-integral at large q behaves as $\int dq \, G_F(q) G_F(q-k)) \simeq \int dq (q^2)^{-2}$. This integral is logarithmically divergent.

We write the integral in the following way (so called external momentum subtraction method)

$$\int dq (G_F(q) G_F(q-k) - G_F(q) G_F(q)) + \int dq \, G_F(q) G_F(q). \qquad (3.58)$$

The first integral is finite. In fact, it is equal

$$\int dq (q^2 - m^2)^{-1}((q-k)^2 - m^2)^{-1} - (q^2 - m^2)^{-1})$$

$$= \int dq ((q-k)^2 - m^2)^{-1}(q^2 - m^2)^{-1}(q^2 - m^2)^{-1}(-(q-k)^2 + q^2). \tag{3.59}$$

The denominator behaves as q^{-6} and nominator $\simeq q$. Hence, d^4q-integral is finite. The second integral in Eq. (3.58) gives an infinite constant (we can make it finite restricting the range of momentum integration at large q). It is important to notice that if we add this infinite second order (in g) term to the term in the 4-point function of the first order (3.53) then we can see that it corresponds just to a replacement of the initial coupling constant $g \to g + C^2 g^2 \int dq G_F(q)^2$ (with a certain positive constant C^2). Hence, if we change the initial coupling constant $g \to g_{bare}$ where

$$g_{bare} = g - C^2 g^2 \int dq G_F(q)^2 \tag{3.60}$$

then the result of the calculation of the four-point correlation function will be finite till the order g^2. Unfortunately, the initial (bare) coupling constant must be negative when the range of q tends to infinity (so the energy is negative). This happens in all theories which are not asymptotically free (only the models containing Yang-Mills fields can be asymptotically free in four dimensions).

We elaborate the mass and coupling constant renormalization in the one loop approximation again in Sect. 7.6 in the framework of the effective action. We still discuss the divergence of Feynman diagrams in $V = g\phi^N$ quantum field theory in d-dimensional space-time in Sect. 3.8. For the general renormalization program see [36, 196, 214].

3.7 Euclidean Correlation Functions

The Feynman propagator satisfies the equation

$$-(\partial_0^2 - \Delta + m^2)\Delta_F(x - y) = \delta(x - y). \tag{3.61}$$

The Fourier transform of the solution of Eq. (3.61) is

$$G_F(p) \equiv \Delta_F(p) = (p_0^2 - \mathbf{p}^2 - m^2)^{-1}. \tag{3.62}$$

If we continue Eqs. (3.61)–(3.62) to the imaginary time and imaginary momenta $x_0 \to ix_0$ and $p_0 \to ip_0$ then $-p_0x_0 + \mathbf{px} \to -p_0x_0 + \mathbf{px}$ and $(-p_0^2 + m^2 + \mathbf{p}^2)^{-1} \to G_E(p) = (p_0^2 + m^2 + \mathbf{p}^2)^{-1}$. We obtain the Euclidean metric. Define $G_E(x_0, \mathbf{x})$ as the Fourier transform of $G_E(p)$ then

$$(-\partial_0^2 - \Delta + m^2)G_E(x - y) \equiv (-\Delta_E + m^2)G_E(x - y) = \delta_E(x - y), \tag{3.63}$$

where $\triangle_E$ denotes the four-dimensional Euclidean Laplacian and δ_E is the δ-function in R^4 (we shall skip E in the δ-function further on). G_E has the Fourier transform

$$G_E(p) = (p_E^2 + m^2)^{-1}, \tag{3.64}$$

where $p_E^2 = p_0^2 + \mathbf{p}^2$. Note that in the Minkowski space $\triangle_F(p)$ had a singularity when $p^2 = m^2$. In such a case the Fourier integrals and momentum integrals in perturbative QFT had to be carefully defined (they exist in the sense of distribution theory [103]). This problem appeared also in the renormalization theory when multiple integrals in momentum space had to be calculated. Physicists noted early that the analytic continuation is a useful technical tool in perturbation expansion as well as when going beyond the perturbation theory [229]. Moreover, in the 70-ties they proved that Euclidean version is completely equivalent to the Minkowski theory , i.e., if we have Euclidean correlation functions then we can construct the vacuum time-ordered correlation functions. From these correlation functions we can get the Hilbert space and the field operators in the Hilbert space (see Sect. 3.10).

Let us continue analytically all free field correlation functions using the formula

$$\begin{aligned} &< 0|T(\phi(x_1).....\phi(x_{2n}))|0 > \\ &= \sum_P < 0|T(\phi(x_{i_1})\phi(x_{i_2}))|0 > ... < 0|T(\phi(x_{i_{2n-1}}).....\phi(x_{i_{2n}}))|0 >, \end{aligned} \tag{3.65}$$

where the sum is over all possible pairings of indices.

We can sum up the correlations for the generating functional (3.33) (after continuation to the imaginary time)

$$Z[f] \equiv \exp\left(-\frac{1}{2}\int dx dy f(x) f(y) G_E(x-y)\right). \tag{3.66}$$

For the development of the Euclidean field theory it was crucial to notice that if we introduce the functional measure (and its rigorous mathematical version of Sect. 1.3)

$$\begin{aligned} d\mu_0 &= d\phi \exp\left(-\frac{1}{2}\int dx (\nabla_E \phi \nabla_E \phi + m^2 \phi^2)\right) \\ &= d\phi \exp\left(-\frac{1}{2}\int dx \phi(-\triangle_E + m^2)\phi\right), \end{aligned} \tag{3.67}$$

where ∇_E is the four-dimensional Euclidean gradient then

$$\exp\left(-\frac{1}{2}\int dx dy f(x) f(y) G_E(x-y)\right) = \int d\mu_0(\phi) \exp(i\int dx f(x)\phi(x))$$

So, $Z[f]$ in Eq. (3.66) is a Fourier transform of a measure.

The integration (3.67) is performed using the Gaussian integral true in arbitrary number of dimensions

$$\int dX \exp(-X\frac{M}{2}X) \exp(JX)\left(\int dX \exp(-X\frac{M}{2}X)\right)^{-1} \tag{3.68}$$
$$= \exp(\tfrac{1}{2}JM^{-1}J)$$

and

$$\left(\int dX \exp\left(-X\frac{M}{2}X\right)\right) = \det\left(\frac{M}{2\pi}\right)^{-\frac{1}{2}}. \tag{3.69}$$

Hence, we have

$$Z[f] = \exp\left(-\frac{1}{2}\int dxdy f(x)f(y)G_E(x-y)\right) = \int d\mu_0 \exp\left(i\int dx f(x)\phi(x)\right). \tag{3.70}$$

As explained in Sect. 1.3 the formula (3.70) for the generating functional $Z[J]$ determines a measure on the set of distributions.

We can continue to the imaginary time also the Gell-Mann-Low formula as follows

$$< 0|T\left(\phi(x_1).....\phi(x_{2n})\exp(-i\int V)\right)|0> \rightarrow$$
$$\int d\mu_0 \exp(-\int dx V(\phi))\phi(x_1).....\phi(x_{2n}) \tag{3.71}$$
$$= \int d\mu_V \phi(x_1).....\phi(x_{2n}),$$

where $d\mu_V$ is discussed in Sect. 1.3. The coupling constant expansion gives the formula for the Euclidean generating functional as an analytic continuation of the Gell-Mann-Low formula (3.38)

$$\sum_{r=0}^{\infty} \int dx_1....dx_r \frac{(-1)^r}{r!} S_r(x_1, ..., x_r) J(x_1)...J(x_r)$$
$$= \left(\int d\mu_0 \exp(-\int V)\right)^{-1} \sum_{r=0}^{\infty} \int dx_1....dx_r \frac{(-1)^r}{r!} J(x_1)...J(x_r) \tag{3.72}$$
$$\times \int d\mu_0 \frac{(-1)^r}{r!}(\int V)^r \phi(x_1)....\phi(x_r)$$

We shall consider in Chap. 5 the functional integral in quantum mechanics from the point of view of the Hamiltonian time evolution. Subsequently, the time will be continued to the imaginary time leading to the Euclidean field theory. The measure $d\mu_V$ with the factor $\exp(-\int dx V)$ after proper regularizations is well-defined and satisfies the positivity condition (1.14). For its definition we need to introduce the volume cutoff and the ultraviolet cutoff restricting the range of momenta so that $V(\phi)$ is a regular function. The cutoffs violate the Euclidean invariance. The aim of the constructive quantum field theory [111] is (after an introduction of mass and charge renormalization counterterms) to show that the cutoffs can be removed leading to a non-trivial Euclidean field theory. Clearly, the positivity condition (1.14) will be satisfied in the limit when we remove cutoffs. As discussed in Sect. 3.10 Euclidean invariant field theory satisfying some additional conditions can be continued analytically in time to quantum field theory. The continuation is on the level of correlation functions. In subsequent chapters we explore the relation between real and imaginary time directly in terms of the Hamiltonian time evolution.

In Chap. 6 a mathematical framework of an analytic continuation (3.71) and of the functional integration for Eq. (3.72) will be discussed in more detail.

3.8 Dimensional Regularization and Power Counting

We write most formulae in four-dimensional space-time. The transition to the d-dimensional space-time or to the Euclidean d dimensional space entails just the replacement of the four (or spatial three) dimensional integrals by d-dimensional (resp. $(d-1)$ space) integrals. So, e.g., the formula for the Feynman propagator in Minkowski space in d-dimensions reads

$$
\begin{aligned}
&i\Delta_F(x-y) \\
&= \tfrac{1}{2}(2\pi)^{-d+1} \int d^{d-1}p\,\omega(\mathbf{p})^{-1} \exp(-i\omega(\mathbf{p})|x_0-y_0|)\exp(i\mathbf{p}(\mathbf{x}-\mathbf{y})) \qquad (3.73)\\
&= (2\pi)^{-d} \int d^d p(-m^2 - \mathbf{p}^2 + p_0^2)^{-1} \exp(ip(x-y)).
\end{aligned}
$$

In Euclidean space

$$
G_E(x-y) = (2\pi)^{-d} \int d^d p(m^2 + p_0^2 + \mathbf{p}^2)^{-1} \exp(ip(x-y)),
$$

where $px = p_0 x_0 + \mathbf{px}$.

It follows from (3.73) that the divergence of the propagator at the coinciding points is increasing with the dimension d. For dimensional regularization (a decrease in d) it is useful to apply the proper time representation

$$
(p_0^2 - \mathbf{p}^2 - m^2)^{-1} = -i \int_0^\infty d\tau \exp(-i\tau(m^2 - p_0^2 + \mathbf{p}^2)).
$$

In such a case the momentum integrals in the Feynman diagrams are Gaussian and can be performed explicitly. The singularity in the momentum space is translated into the power-law singularity of the proper time integrals. As an example the integral (3.57) arising in ϕ^4 theory in the computation in momentum space of the Feynman diagram with one closed loop, four external lines, two vertices and two internal lines in d-dimensions is

$$
\begin{aligned}
&\int d^d q \int_0^\infty \int_0^\infty d\tau_1 \int d\tau_2 \exp(-i(\tau_1+\tau_2)m^2 + i\tau_2 k^2) \\
&\times \exp i(\tau_1 q^2 + \tau_2(q^2 - 2qk)) \simeq \int_0^\infty d\tau_1 \int_0^\infty d\tau_2(\tau_1+\tau_2)^{-\frac{d}{2}}.
\end{aligned} \qquad (3.74)
$$

In Eq. (3.74) the diagrams when expressed by τ-integrals can be considered as functions of a continuous variable d. When $d \to 4$ the integral becomes logarithmically divergent at small τ. The procedure (3.58) of a subtraction of the external momentum is equivalent to a subtraction of the divergent $k = 0$ τ-integrals in Eq. (3.74). We say that the power of divergence ν of the integral (3.74) in the proper time at small τ is $\nu = \frac{d}{2} - 2$. This order of divergence would be the same if we worked in momentum space in Euclidean space-time and used the coordinates $\xi = |\mathbf{q}_E|^2$. Then, the divergence of the integral (3.57) at large ξ will be $\int \frac{d\xi}{\xi} \xi^{\frac{d}{2} - 2}$.

On the basis of the Feynman rules in momentum space formulated in Sect. 3.6 we can estimate the order of divergence ν of each Feynman diagram [209]. We consider a diagram in ϕ^N quantum field theory with E external lines, I internal lines, $\mathcal{V}$ vertices and L loops. In each loop of the diagram there is a d-dimensional integral and p^{-2} propagator for an internal line. Hence, the power of divergence of such a diagram, is

$$\nu = Ld - 2I, \tag{3.75}$$

where $2I$ comes out because of $(p^2)^{-I}$ from propagators. According to the Feynman rules of Sect. 3.6 in each vertex there is a conservation of momenta expressed by the δ-function. So the number of loops, equal to the number of momenta for integration, is

$$L = I - \mathcal{V} + 1. \tag{3.76}$$

To each $\mathcal{V}$ vertex there enter N-lines. Hence at the $\mathcal{V}$th order of perturbation expansion the number of vertices is related to the number of external lines E and internal lines (the same internal line is connected to two vertices) as

$$N\mathcal{V} = E + 2I. \tag{3.77}$$

Solving Eqs. (3.75)–(3.77) we obtain for the index of divergence ν at the $\mathcal{V}$th order of perturbation expansion

$$\nu = d - E\left(\frac{d}{2} - 1\right) + \mathcal{V}\left(N\left(\frac{d}{2} - 1\right) - d\right). \tag{3.78}$$

From Eq. (3.78) it follows that the theories with $N \geq 5$ are not renormalizable in $d = 4$ dimensions because the index of divergence ν is increasing with the order $\mathcal{V}$ of calculations. For $N = 4$ and $d = 4$ we have $\nu = 4 - E$ as checked in Sect. 3.6 for $E = 4$. In the same model if $E = 2$ then $\nu = 2$ (quadratic divergence of the mass renormalization) as will be obtained in Sect. 7.6.

We shall still discuss the one loop diagrams in Sect. 7.6 confirming the formula (3.78). We encounter the subtraction of the proper-time integrals and a renormalization of such integrals in the calculation of the effective action by means of the heat kernel in Chap. 7.

3.9 Generating Functional: A Perturbative Formula

Using the functional integration formula of Sect. 3.7 we can show that the generating functional (3.33) (Euclidean version (3.72), in the Euclidean formulation we change the current $J \to iJ$) has the representation

$$
\begin{aligned}
Z_E[J] &= \int d\phi \exp(-\tfrac{1}{2}\int dx (\partial_\mu \phi \partial_\mu \phi + m^2 \phi^2)) \exp(-g \int dx V(\phi) - \int dx J\phi) \\
&= \exp\left(-g \int dx V(-\tfrac{\delta}{\delta J(x)})\right) \int d\phi \exp(-\int dx \mathcal{L}_0(\phi) - \int dx J\phi) \\
&= \exp\left(-g \int dx V(-\tfrac{\delta}{\delta J(x)})\right) \exp(\tfrac{1}{2} J G_E J),
\end{aligned}
$$

(3.79)

where

$$
\mathcal{L}_0 = \frac{1}{2}\partial_\mu \phi \partial_\mu \phi + \frac{m^2}{2}\phi^2.
$$

In order to prove Eq. (3.79) we note that the action of the functional differentiation $-\tfrac{\delta}{\delta J(x)}$ applied to $\exp(-\int J\phi)$ in the second line of Eq. (3.79) is replacing $V(-\tfrac{\delta}{\delta J(x)})$ by $V(\phi)$ under the integral sign giving (after a summation of n-point functions) the functional integration formula (3.72) for the generating functional. When $\exp\left(-g \int dx V(-\tfrac{\delta}{\delta J(x)})\right)$ is outside the functional integral in the second line of Eq. (3.79) then we can explicitly perform the functional integral arriving at the formula in the third line.

The formula for the generating functional is useful for a derivation of the Feynman rules together with proper combinatorial factors which follow from functional differentiation. For the combinatorics in ϕ^4 it is sufficient to calculate the integral

$$
\int dx \exp\left(-\frac{1}{2}ax^2\right)\exp(jx)\left(\int dx \exp(-\frac{1}{2}ax^2)\right)^{-1} = Z(j) = \exp\left(\frac{1}{2}a^{-1}j^2\right).
$$

(3.80)

Then

$$
\int dx \exp\left(-\frac{1}{2}ax^2\right)x^4\left(\int dx \exp\left(-\frac{1}{2}ax^2\right)\right)^{-1} = \partial_j^4 \exp\left(\frac{1}{2}a^{-1}j^2\right)\Big|_{j=0} = 3a^{-2}.
$$

For ϕ^{2n}

$$
\int dx \exp\left(-\frac{1}{2}ax^2\right)x^{2n}\left(\int dx \exp\left(-\frac{1}{2}ax^2\right)\right)^{-1} = 2^{-n}a^{-n}(2n)!(n!)^{-1}. \quad (3.81)
$$

These formulas show the number of terms in Feynman diagrams. An estimate of the number of terms shows that the perturbation series is not convergent in the ϕ^{2n} model for $n \geq 2$ (it is at most an asymptotic series; for more detailed estimates see [177]). In Eq. (3.79) we should take into account the Euclidean normal ordering according to the formula

$$
: \exp(\lambda \phi) := \exp(\lambda \phi) < \exp(\lambda \phi) >^{-1}
$$

Then, on the rhs of Eq. (3.79) some differentials are omitted, e.g., in : ϕ^4 : the operator $(\frac{\delta}{\delta J(x)})^4$ does not act upon $(JGJ)^2$ (in general on any polynomials of J producing $G(0)$). This follows from the fact that $Z_E(J)$ is a sum of the perturbation series (3.33) of vacuum correlation functions. These correlation functions (by definition of the normal ordering) do not contain the $G(0)$ terms (see Eqs. (3.15)–(3.16)).

3.10   The Euclidean Quantum Field Theory: Osterwalder-Schrader Formulation

So far we have discussed the analytic continuation of quantum field theory from the Minkowski space-time to Euclidean space on a formal level of perturbation theory. The relation between non-commutative fields in the Minkowski space-time and commutative Euclidean fields has been studied by Osterwalder and Schrader [191]. The result is that the correlation functions of the relativistic quantum field theory in Minkowski space satisfying the Wightman axioms [228] (known to have an extension to the complex domain owing to the spectral condition) can be analytically continued to the Euclidean points corresponding to the Euclidean metric. The resulting correlation functions (Schwinger functions S_n^E) are symmetric under permutation of the Euclidean points. They define the commutative Euclidean field theory with some particular properties (axioms). The main achievement of Osterwalder and Schrader [191] is the proof of the inverse theorem: from Euclidean theory we can construct the quantum field theory in Minkowski space-time. The crucial property, allowing the reconstruction, is the Osterwalder-Schrader (OS) positivity property of the Euclidean generating functional. Let us consider real functions $f(x_0, \mathbf{x}) \in S$ (Sect. 1.2) with their support on $x_0 \geq 0$. The generating functional is defined by Schwinger functions as (we write $J = -if$ in Eq. (3.79) where f is real)

$$Z[f] = 1 + \sum_{n=1}^{\infty} \frac{i^n}{n!} \int dx_1....dx_n S_n^E(x_1, \ldots, x_n) f(x_1)....f(x_n).$$

The OS positivity is defined in terms of the generating functional as the inequality

$$\sum_{jk} \lambda_k^* \lambda_j Z[f_j - \theta f_k] \geq 0, \tag{3.82}$$

where $\theta f(x) = f(\vartheta x)$ with $\vartheta(x_0, \mathbf{x}) = (-x_0, \mathbf{x})$.

In terms of $Z[f]$ we can construct a scalar product on the set of functions of the form

$$F = \sum_j c_j \exp(if_j), \tag{3.83}$$

where c_j are complex numbers and f_j are real and have their support in the half-space $x_0 \geq 0$. We define

$$\left(\exp(if_2), \exp(if_1) \right) = Z[f_1 - \theta f_2] \tag{3.84}$$

and extend this definition through linearity to all functions of the form (3.83). From (3.82) it follows that such a scalar product is positive definite. By completion of the set of functions in the norm (3.84) we obtain a Hilbert space.

We can define a semigroup T_t as

$$T_t \exp(if(x_0, x_1, ..., x_{d-1}) = \exp(if(x_0 - t, x_1, ..., x_{d-1})). \tag{3.85}$$

It can be proved that with the scalar product (3.84) the semigroup T_t is generated by an operator $H \geq 0$, i.e., $T_t = \exp(-Ht)$. In this way the Hilbert space and the physical Hamiltonian H are constructed. Then, the invariance with respect to the Lorentz group follows from the Euclidean invariance. By means of the positive definite Hamiltonian we can continue the Euclidean fields to the Minkowski space-time and construct quantum fields from the time-zero fields as in Eq. (3.1).

We illustrate the Osterwalder-Schrader methods on the example of the free field (a similar argument applies to generalized free fields when in the Euclidean two-point function (3.64) we have an integral over m with a positive integration measure [149]). For this purpose we use the integral

$$\int\limits_{-\infty}^{\infty} dk \exp(ikt)(k^2 + \omega^2)^{-1} = \frac{\pi}{\omega} \exp(-\omega|t|). \tag{3.86}$$

For the free field we have

$$Z[f] = \exp\left(-\frac{1}{2} f G_E f \right), \tag{3.87}$$

where G_E is defined in Eqs. (3.63)–(3.64). Then, for $x_0, y_0 \geq 0$

$$\begin{aligned} G_E(x - \vartheta y) &= (2\pi)^{-4} \int d^4 p \exp(ip(x - \vartheta y))(p^2 + m^2)^{-1} \\ &= \frac{1}{2}(2\pi)^{-3} \int d^3 p \exp(i\mathbf{p}(\mathbf{x} - \mathbf{y}))\omega^{-1} \exp(-(x_0 + y_0)\omega(p)). \end{aligned} \tag{3.88}$$

The OS positivity condition is satisfied because we can apply the positivity argument (1.14) of Sect. 1.3. $G_E(x - \vartheta y)$ is positive definite in the sense

$$\sum c_i c_k^* G_{ik} \equiv \sum c_i c_k^* \int dx_0 d\mathbf{x} dy_0 d\mathbf{y}\, f_i(x_0, \mathbf{x}) f_k(y_0, \mathbf{y}) G(x - \vartheta y) \geq 0,$$

where we denoted $G_{ik} = \int dx_0 d\mathbf{x} dy_0 d\mathbf{y}\, f_i(x_0, \mathbf{x}) f_k(y_0, \mathbf{y}) G(x - \vartheta y)$. This inequality easily follows from Eq. (3.88). Hence,

$$\sum c_i c_k^* Z[f_i - \theta f_k] = \sum \left(c_i \exp(-\tfrac{1}{2} G_{ii}) \right) \left(c_k \exp(-\tfrac{1}{2} G_{kk}) \right)^* \exp(G_{ik}) \geq 0 \tag{3.89}$$

on the basis of the statement (1.16) of Sect. 1.3.

Then, consider T_t of Eq. (3.85)

$$(\exp(if), T_t \exp(if)) = Z[f' - \theta f], \tag{3.90}$$

where $f'(x_0, \mathbf{x}) = f(x_0 - t, \mathbf{x})$. Hence,

$$\begin{aligned}
(\exp(if), T_t \exp(if) &= Z[f' - \theta f] = Z[f]^2 \\
&\times \exp\left(-(2\pi)^{-3} \int dx_0 d\mathbf{x} \int dy_0 dy d\mathbf{p} \exp(i\mathbf{p}(\mathbf{x} - \mathbf{y})) \right. \\
&\times \omega^{-1} \exp(-(x_0 + y_0 + t)\omega(p)) f(x_0, \mathbf{x}) f(y_0, \mathbf{y}) \Big).
\end{aligned} \tag{3.91}$$

From Eqs. (3.90)–(3.91) we can conclude that T_t is a semigroup $T_t = \exp(-Ht)$, where H is self-adjoint and $H \geq 0$. For time-zero quantum fields of Sect. 2.3 we have

$$\begin{aligned}
< 0|\phi(h) T_t \phi(h)|0 > &= \tfrac{1}{2} (2\pi)^{-3} \int d\mathbf{x} d\mathbf{y} d\mathbf{p} \exp(i\mathbf{p}(\mathbf{x} - \mathbf{y})) \omega^{-1} \\
&\times \exp(-t\omega(\mathbf{p})) h(\mathbf{x}) h(\mathbf{y})
\end{aligned} \tag{3.92}$$

We obtain Eq. (3.92) calculating the action of $\exp(-tH)$ (with H of Eq. (2.26)) upon the states (2.36) generated by the action of the free field on the Fock vacuum of Chap. 2, Eq. (3.92) is in agreement with Eq. (3.91) if $f(x_0, \mathbf{x}) = \delta(x_0) h(\mathbf{x})$. It follows that $T_t = \exp(-\omega(\mathbf{p})t)$ when acting on $a^+(\mathbf{p})|0 >$.

A local Lagrangiann can be split into the positive time half-space E_+ and negative time half-space θE_+

$$\int dx \mathcal{L}(\phi) = \int_{E_+} dx \mathcal{L}(\phi) + \theta \int_{E_+} dx \mathcal{L}(\phi), \tag{3.93}$$

where E_+ is the half-space with $x_0 \geq 0$. Then, in a heuristic expression for the generating functional

$$Z[f] = \int d\phi \exp\left(-\int dx \mathcal{L} + i \int dx \phi f \right) \tag{3.94}$$

$Z[f]$ satisfies the OS positivity as

$$\begin{aligned}
\sum_{j,k} c_j^* c_k Z[f_j - \theta f_k] &= \left(\int d\phi \sum_j c_j \exp(-i\phi(f_j)) \exp(-\int_{E_+} dx \mathcal{L}(\phi)) \right)^* \\
&\times \left(\int d\phi \sum_j c_j \exp(-i\phi(\theta f_j)) \exp(-\theta \int_{E_+} dx \mathcal{L}(\phi)) \right) \geq 0,
\end{aligned} \tag{3.95}$$

because the expression (3.95) is of the form F^*F as the integrals on both half spaces are the same. The problem with OS positivity in formal perturbative QFT is that QFT requires regularization which usually breaks the locality. For Lagrangians with the first order derivatives the lattice is the exceptional regularization which preserves the OS positivity. Then, the formal expressions (3.94)–(3.95) have a precise meaning on the lattice as will be discussed in Chap. 12. An addition to the Lagrangian of terms of higher order in derivatives can serve as a regularization. The heuristic argument (3.93) still works. However, in the lattice approximation there appear terms mixing the E_+ and E_- half-spaces which in the models with higher derivatives violate the OS positivity. This shows that heuristic arguments sometimes have fatal defects as it is known that the theories with higher derivatives are not OS positive and have no realization as QFT with positive energy.

3.11 Heisenberg Picture: The Asymptotic Fields

We can approach quantization in the Heisenberg framework by solving differential equations with the boundary condition saying that asymptotically the quantum field behaves as the free field [246]. We assume that interacting fields satisfy (after renormalization) the same differential equations as the classical fields. We wish to solve

$$\partial_\mu \partial^\mu \phi + m^2 \phi = -V'(\phi). \tag{3.96}$$

Let us introduce

$$\Delta_{ret}(x) = \theta(x_0)\Delta(x), \tag{3.97}$$

where

$$\Delta(x) = \Delta^{(+)}(x) - \Delta^{(+)}(-x) \tag{3.98}$$

is the commutator of free fields. We have by direct differentiation

$$(\partial_\mu \partial^\mu + m^2)\Delta_{ret} = -i\delta(x). \tag{3.99}$$

We can rewrite Eq. (3.96) as an integral equation

$$\phi(x) = \phi_{in}(x) - i\int dy \Delta_{ret}(x-y)V'(\phi(y)), \tag{3.100}$$

where ϕ_{in} is a free field.

Using the advanced Green function $\Delta_{adv}(x) = \theta(-x_0)\Delta(x)$ we can write Eq. (3.96) as

$$\phi(x) = \phi_{out}(x) + i\int dy \Delta_{adv}(x-y)V'(\phi(y)), \tag{3.101}$$

where ϕ_{out} is another free field.

When $x_0 \to -\infty$ then in Eq. (3.100) $\phi \to \phi_{in}$ because $\Delta_{ret}(x-y) \to 0$ when $x_0 \to -\infty$. This is the same limit as in the interaction picture where we have chosen ϕ_{in} as an initial condition at $x_0 = -\infty$. Moreover, when $x_0 \to +\infty$ then $\Delta_{ret}(x-y) \to \Delta(x-y)$. As

$$(\partial_\mu \partial^\mu + m^2)\Delta(x) = 0 \tag{3.102}$$

then for $x_0 \to +\infty$ we have from Eq. (3.100) that

$$(\partial_\mu \partial^\mu + m^2)\phi(x) = 0. \tag{3.103}$$

Hence, $\phi(x)$ when $x_0 \to +\infty$ tends to the free field. This free field of Eq. (3.100) at $x_0 = +\infty$ must be ϕ_{out} of Eq. (3.101) as for $x_0 \to +\infty$ in Eq. (3.101) $\phi \to \phi_{out}$. Similarly, if we take the limit $x_0 \to -\infty$ in Eq. (3.101) then in this limit we obtain the free field which must be ϕ_{in}. So, Eqs. (3.100) and (3.101) are consistent. We can define the S-matrix by means of the equation (together with the assumption that S leaves the physical vacuum invariant)

$$\phi_{out} = S^{-1}\phi_{in}S. \tag{3.104}$$

S is an evolution operator from $-\infty$ to $+\infty$ so one can claim that this is the same operator as the one defined in the interaction picture. In fact from Eqs. (3.100)–(3.101) and (3.104) one can derive [246]

$$S = T\left(\exp\left(-i\int_{-\infty}^{\infty} V(\phi_{in})dx\right)\right) \tag{3.105}$$

in agreement with Eq. (3.10) of Sect. 3.11. The rather formal arguments of this section will be formulated as a starting point of QFT (axioms) in the formulation of Lehmann, Symanzik and Zimmermann (LSZ [175]) in the next section.

3.12 Reduction Formulas

We have explained in a heuristic way that the interacting fields ϕ in the interaction picture can be defined in such a way that
for $x_0 \to -\infty$ we have $\phi \to \phi_{in}$
and
for $x_0 \to +\infty$ asymptotically $\phi \to \phi_{out}$
where
ϕ_{in} and ϕ_{out} are free fields.
Moreover, we defined the operator S (scattering matrix) as

$$\phi_{out} = S^+\phi_{in}S$$

with the stability condition

$$S|\Phi_0> = |\Phi_0>, \tag{3.106}$$

where $|\Phi_0>$ is the true physical vacuum which is the same for the free fields ϕ_{in} as well as ϕ_{out}. S is an unitary evolution operator from $-\infty$ to $+\infty$. Then, the scattering amplitude is

$$
\begin{aligned}
&< \Phi_0|a_{in}(p_1)...a_{in}(p_k)Sa_{in}^+(q_1)....a_{in}^+(q_r)|\Phi_0> \\
&=< \Phi_0|a_{out}(p_1)...a_{out}(p_k)a_{in}^+(q_1)....a_{in}^+(q_r)|\Phi_0>,
\end{aligned}
\tag{3.107}
$$

This framework constitutes the foundation of the LSZ (axiomatic) approach to quantum field theory formulated by Lehmann et al. [175]. The assumption is that there is a local relativistic quantum field $\phi(x)$ which tends to free fields ϕ_{out} (ϕ_{in}) in the remote future (resp.past). Then, LSZ have shown that the S matrix defined by Eq. (3.104) can be expressed by time ordered vacuum (the physical vacuum) correlation functions of $\phi(x)$. In general, we are unable to construct the quantum field ϕ nonperturbatively. However, in the interaction picture in Sect. 3.3 we have obtained a formula for quantum fields and their vacuum correlation functions (Gell-Mann-Low formula) which can be calculated by means of a perturbation expansion (Feynman diagrams). It is known (Haag's theorem [116, 131]) that the formulas of the interaction picture representing the relativistic interacting field in terms of the free field have no rigorous mathematical sense. However, the formula for the scattering amplitude derived in the interaction picture is satisfied in any perturbative approach to quantum field theory. In quantum electrodynamics the perturbative scattering amplitudes are in excellent agreement with experiments.

We outline now the LSZ derivation of the formula for the scattering amplitude in terms of time-ordered correlation functions of quantum fields. We express the creation operators in (3.107) first by the free field and next by an interacting field. Using the expansion of the free field in creation and annihilation operators we can obtain from Eq. (2.22)

$$a_{in}^+(q) = i \int d\mathbf{x}(\phi_{in}(x)\partial_0 f_q - f_q\partial_0\phi_{in}(x)). \tag{3.108}$$

Then, for any function F

$$\int d\mathbf{x}F(x_0 = -\infty) = \int d\mathbf{x}F(x_0 = +\infty) - \int_{-\infty}^{\infty} dx_0 d\mathbf{x}\partial_0 F(x_0). \tag{3.109}$$

From our assumptions on the asymptotic limit and Eq. (3.108) we have

$$a_{in}^+(q) = \lim_{x_0 \to -\infty} i \int d\mathbf{x}(\phi(x)\partial_0 f_q - f_q\partial_0\phi(x)). \tag{3.110}$$

We insert the limit on the rhs of Eq. (3.110) into the lhs of Eq. (3.109) and subsequently in the scattering amplitude (3.107) (beginning with q_1). The first term on the rhs of Eq. (3.109) will give a_{out}^+. We omit such a term as it is describing the situation when the q_1 particle does not scatter at all (so we have a scattering amplitude for $r - 1$ particles). There remains the second term on the rhs of Eq. (3.109) which is of the form

$$- i \int dx_0 d\mathbf{x} \partial_0 (\phi(x) \partial_0 f_q - f_q \partial_0 \phi(x)). \tag{3.111}$$

We use the fact that f_q satisfies the Klein-Gordon equation. Hence,

$$\partial_0^2 f_q = \Delta f_q + m^2 f_q. \tag{3.112}$$

The differentiation over x_0 in Eq. (3.111) gives the rhs of Eq. (3.112) times ϕ when acting on f_q. The first order derivative terms cancel in Eq. (3.111), whereas the term $f_q \partial_0^2 \phi$ can be combined with the term (3.112) (multiplied by ϕ in Eq. (3.111) and subsequently integrated by parts over spatial coordinates) to give in Eq. (3.111)

$$i \int dx f_q (\partial_\mu \partial^\mu + m^2) \phi(x) \equiv i \int dx f_q K G_x \phi. \tag{3.113}$$

We apply this procedure first to $a_{in}^+(q_1)$ in order to express it by Eq. (3.113). Then, we repeat it in application to $a_{in}^+(q_2)$. We express the product $a_{in}^+(q_1) a_{in}^+(q_2)$ by

$$- \int dx dx' f_{q_1} f_{q_2} T \left((\partial_\mu \partial^\mu + m^2) \phi(x) (\partial_\mu \partial^\mu + m^2) \phi(x') \right) \tag{3.114}$$

The time x_0' for $a_{in}^+(q_2)$ is at $-\infty$ (because of Eq. (3.110)). So it will always be earlier than x. Hence, we have replaced $\phi(x) \phi(x')$ by $T(\phi(x) \phi(x'))$ in Eq. (3.114). It can be seen that repeating the procedure in Eq. (3.114) for the a_{in}^+ operators we obtain a time-ordered product of interacting fields. We can continue this method applying it to the $a_{out}(p)$ fields in Eq. (3.107) using the formula

$$a_{out}(p) = \lim_{y_0 \to \infty} i \int d\mathbf{y} (f_p^* \partial_0 \phi(y) - \phi(y) \partial_0 f_p^*) \tag{3.115}$$

We use the rhs of Eq. (3.109) in order to express the limit $y_0 \to +\infty$ in Eq. (3.115) by a derivative of the interacting field. Now, the time will be bigger than the time inside $T(...)$ hence we can move $\phi(y)$ inside $T(...)$ and the result will be of the form (3.114) (with a conjugation of f because of f^* in Eq. (3.115)). As a result we get $f_p^*(y) f_{q_1}(x) f_{q_2}(x') K G_y K G_x K G_{x'} T(\phi(y) \phi(x) \phi(x'))$. Continuing the procedure we arrive at the formula

$$< p_1, ..., p_k|S|q_1, ...q_r >= i^{k+r} \int dy_1..dy_k dx_1...dx_r$$
$$f_{p_1}^*(y_1)....f_{p_k}^*(y_k)f_{q_i}(x_1)....f_{q_r}(x_r)$$
$$KG_{y_1}....KG_{y_k}KG_{x_1}....KG_{x_r} < \Phi_0|T\Big(\phi(y_1)...\phi(y_k)\phi(x_1)...\phi(x_r)\Big)|\Phi_0 >,$$

$$\tag{3.116}$$

where

$$KG = \partial_\mu \partial^\mu + m^2 \tag{3.117}$$

is the Klein-Gordon operator.

In Sect. 3.3 we have expressed the vacuum expectation values of time-ordered products of interacting fields in the physical vacuum by expectation values of free fields in the Fock vacuum (Gell-Man-Low formula (3.37)). In this way we obtain Feynman diagrams for scattering amplitudes from Eq. (3.116).

As an example consider the model discussed in Sect. 3.3 $V = g : \phi^4 :$. We compute the scattering amplitude $q_1 + q_2 \rightarrow p_1 + p_2$. We calculate the time-ordered correlation functions using the Gell-Mann-Low formula. In the lowest order we obtain as in (3.39)

$$-ig \int dx \Delta_F(x - x_1)\Delta_F(x - x_2)\Delta_F(x - y_1)\Delta_F(x - y_2). \tag{3.118}$$

We insert this expression in the reduction formula (3.116). KG_{x_1} acting upon $\Delta_F(x - x_1)$ gives $-i\delta(x - x_1)$. The integration of the functions f_q and f_p^* over x_j and y_k in Eq. (3.116) leads to the δ-function describing a conservation of momenta. As a result the scattering amplitude in the lowest order of perturbation expansion is

$$-ig\delta(p_1 + p_2 - q_1 - q_2). \tag{3.119}$$

We recognize that the perturbative scattering amplitude (3.116) is given by the same formula (3.53) as the Feynman diagram rule of Sect. 3.6 for the time-ordered correlation functions in the momentum space except that the external lines are removed by the action of the KG operators (amputated Feynman diagrams) . The relation between the Feynman diagrams in momentum space and scattering amplitudes is discussed in [21, 80].

3.13 Exercises

3.1 Prove Eqs. (3.15)–(3.16) and calculate the coefficients c_n
   Hint: the calculation can be reduced to differentiation over λ of the exponentials (3.14).
3.2 From the definition of the S-matrix in terms of $U_I(\infty, -\infty)$ calculate the amplitude (3.11) for the scattering $p_1 + p_2 \rightarrow p_3 + p_4$ in the lowest order of $g\phi^4$ interaction (with the normal ordering of the interaction).

3.3 Calculate the Fourier transform of $G_F(p)$ as explained below Eq. (3.58).

3.4 Using the differential formula in Sect. 3.9 for the generating functional in the model $g\phi^4$ calculate to lowest orders

$$< \Phi_0|T\Big(\phi(x_1)\phi(x_2)\phi(x_3)\phi(x_4)\Big)|\Phi_0 >$$

Check that the Feynman rules follow from differentiation rules.

3.5 Show that the renormalization of the scattering amplitude in Sect. 3.6 at order g^2 can be absorbed in the coupling constant redefinition (3.60) (calculate the constant C).

3.6 Perform the transition to an imaginary time at each order of the Gell-Mann-Low formula as indicated in Eq. (3.72).

3.7 Show Eq. (3.108).

3.8 Calculate the scattering amplitude $p_1 + p_2 \to p_3 + p_4$ till the order g^2 in $g\phi^4$ model using the reduction formula (3.116).

Chapter 4
Thermal States and Quantum Scalar Field on a Curved Manifold

Abstract In this chapter we discuss some contemporary extensions of quantum field theory to mixed states and to curved manifolds. There can be a connection between these two topics as quantum fields in some moving frames (accelerated frames) can be viewed as thermal fields (the Unruh effect) and quantum fields on a black hole background have the thermal spectrum. We derive the formula for the correlation functions of the scalar quantum field in the quantum Gibbs (thermal) state. Then, we approach the problem of a quantization of the scalar field defined on a manifold. We formulate canonical commutation rules for quantization. We solve the heuristic eigenvalue equation to find a Gaussian ground state in a static metric. The ground state wave function is used to construct a Gaussian measure which defines the Hilbert space of square integrable functions (the Schrödinger representation of QFT).

In this chapter we discuss some contemporary extensions of quantum field theory to mixed states and to curved manifolds. There can be a connection between these two topics as quantum fields in some moving frames (accelerated frames) can be viewed as thermal fields (the Unruh effect [236]) and quantum fields on a black hole background have the thermal spectrum [107]. Quantum field theory in a thermal state and on a manifold finds applications to the theory of quark matter in heavy ion collisions, to neutron stars and to a description of the hot stage of the universe at the Big Bang. In Sect. 4.1 we derive the formula for the correlation functions of the scalar quantum field in the quantum Gibbs (thermal) state. These correlation functions will be applied in Chap. 9 to photons and in Chap. 10 to gravitons. In Sect. 4.2 we approach the problem of a scalar field quantization on the globally hyperbolic manifold. We formulate canonical commutation rules for quantization. We solve the heuristic eigenvalue equation to find the ground state in a static metric. The ground state wave function ψ^g determines the Gaussian measure $d\mu_0 = |\psi^g|^2 d\phi$ which defines the Hilbert space for quantum field theory. We discuss the Wigner function of the ground state which allows to calculate the ground state correlation functions in free field theory.

© The Author(s), under exclusive license to Springer Nature Switzerland AG 2023
Z. Haba, *Lectures on Quantum Field Theory and Functional Integration*,
https://doi.org/10.1007/978-3-031-30712-6_4

4.1   Fields at Finite Temperature

Till now we have considered quantum fields in pure states. The contemporary applications ,e.g., in the heavy ion collisions [168], treat scattering of relativistic particles in a heat bath of other particles. For a description of such processes it is necessary to extend the formalism of QFT from pure states to Gibbs states at temperature $T \equiv k_B^{-1}\beta^{-1}$ (where k_B is the Boltzmann constant). Let us consider the free Hamiltonian modified by the chemical potential μ. The chemical potential is needed to take into account the condition that the number of particles is preserved in systems with an arbitrary number of particles described by the grand canonical ensemble (the number of photons is not preserved, hence $\mu = 0$ for photons in Chap. 9). Now,

$$H = \int \frac{d\mathbf{p}}{2\omega(\mathbf{p})}(\omega - \mu)a^+(\mathbf{p})a(\mathbf{p}). \tag{4.1}$$

We can show by means of the Campbell-Baker formula (or by differentiating the lhs of Eq. (4.2) over β and solving the resulting equation) that

$$\exp(-\beta H)a\exp(\beta H) = \exp(\beta\hbar(\omega - \mu))a, \tag{4.2}$$

$$\exp(-\beta H)a^+ \exp(\beta H) = \exp(-\beta\hbar(\omega - \mu))a^+. \tag{4.3}$$

Equations (4.2)–(4.3) constitute just the imaginary time version of the unitary evolution of creation-annihilation operators.

In quantum statistical mechanics we define an expectation value of an observable F in the thermal state as

$$\langle F\rangle_\beta = Z_\beta^{-1}Tr(\exp(-\beta H)F) \tag{4.4}$$

with

$$Z_\beta = Tr(\exp(-\beta H)). \tag{4.5}$$

Let us calculate (using $Tr(AB) = Tr(BA)$ and Eq. (4.2))

$$\begin{aligned}
\langle a(\mathbf{p}')^+ a(\mathbf{p})\rangle_\beta &= Z_\beta^{-1}Tr\Big(a(\mathbf{p})\exp(-\beta H)a(\mathbf{p}')^+\Big) \\
&= Z_\beta^{-1}Tr\Big(\exp(-\beta H)\exp(\beta H)a(\mathbf{p})\exp(-\beta H)a(\mathbf{p}')^+\Big) \\
&= \exp(-\beta\hbar(\omega(\mathbf{p}) - \mu))Z_\beta^{-1}Tr\Big(\exp(-\beta H)a(\mathbf{p})a(\mathbf{p}')^+\Big).
\end{aligned} \tag{4.6}$$

We apply the commutation relation

$$[a(\mathbf{p}), a^+(\mathbf{q})] = 2\omega(\mathbf{p})\hbar\delta(\mathbf{p} - \mathbf{q}) \tag{4.7}$$

in order to rewrite aa^+ as a^+a plus δ-function. We obtain from Eq. (4.6)

$$\langle a(\mathbf{p}')^+ a(\mathbf{p}) \rangle_\beta = \exp(-\beta\hbar(\omega(\mathbf{p}) - \mu))\langle a(\mathbf{p}')^+ a(\mathbf{p}) \rangle_\beta$$
$$+ 2\omega(\mathbf{p})\hbar\delta(\mathbf{p} - \mathbf{p}') \exp(-\beta\hbar(\omega(\mathbf{p}) - \mu)). \tag{4.8}$$

Hence,

$$\langle a(\mathbf{p}')^+ a(\mathbf{p}) \rangle_\beta = 2\omega(\mathbf{p})\hbar\delta(\mathbf{p} - \mathbf{p}')\Big(\exp(\beta\hbar(\omega(\mathbf{p}) - \mu)) - 1 \Big)^{-1}. \tag{4.9}$$

This is the Bose-Einstein distribution.

Let us now calculate the field correlation functions. We have from Sect. 2.2

$$\phi(x) = \int \frac{d\mathbf{p}}{2\omega(\mathbf{p})}\Big(a(\mathbf{p}) f_p(x) + a^*(\mathbf{p}) f_p^*(x)\Big), \tag{4.10}$$

where

$$f_p(x) = (2\pi)^{-\frac{3}{2}} \exp(-ipx) \tag{4.11}$$

with $px = p_0 x_0 - \mathbf{p}\mathbf{x}$ and $p_0 \equiv \omega(\mathbf{p}) = \sqrt{\mathbf{p}^2 + m^2}$.

The calculation of the field correlation functions is reduced to the one of expectation values of creation and annihilation operators. It can easily be seen that $< aa >_\beta = 0$ and $< a^+ a^+ >_\beta = 0$ because the trace is a sum of expectation values over n-particle states. The product aa decreases the number of particles hence the diagonal over n-particle states is zero (the same argument applies for creation operators). Hence, in the product of $\phi\phi$ only the creation-annihilation terms are non-zero.

We have

$$< \phi(x)\phi(y) >_\beta = (2\pi)^{-3} \int \frac{d\mathbf{p}}{2\omega(\mathbf{p})} \int \frac{d\mathbf{q}}{2\omega(\mathbf{q})}$$
$$\Big(\exp(ipx - iqy) < a^+(\mathbf{p})a(\mathbf{q}) >_\beta + \exp(-ipx + iqy) < a(\mathbf{p})a^+(\mathbf{q}) >_\beta \Big). \tag{4.12}$$

We can transform the second term in Eq. (4.12) to the form of the first term using the commutation relation (4.7). The expectation value of the first term has been calculated in Eq. (4.9). From Eq. (4.12) we obtain

$$< \phi(x)\phi(y) >_\beta = \Delta^{(+)}(x - y) + 2\hbar(2\pi)^{-3} \int \frac{d\mathbf{p}}{2\omega(\mathbf{p})}$$
$$\times \Big(\exp(\beta\hbar(\omega(\mathbf{p}) - \mu)) - 1 \Big)^{-1} \cos(\omega(\mathbf{p})(x_0 - y_0)) \cos(\mathbf{p}(\mathbf{x} - \mathbf{y}))$$

with

$$\Delta^{(+)}(x - y) = < 0|\phi(x)\phi(y)|0 > = \hbar(2\pi)^{-3} \int \frac{d\mathbf{p}}{2\omega(\mathbf{p})} \exp(-ip(x - y)),$$

where we write $\hbar$ explicitly. In order to exhibit the behavior at $\hbar \rightarrow 0$ we rewrite the formula for the two-point function as

$$
\begin{aligned}
&< \phi(x)\phi(y) >_\beta = \hbar(2\pi)^{-3} \int \frac{d\mathbf{p}}{2\omega(\mathbf{p})} \cos(\mathbf{p}(\mathbf{x} - \mathbf{y})) \\
&\left(\coth(\tfrac{1}{2}\beta\hbar(\omega(\mathbf{p}) - \mu)) \cos(\omega(\mathbf{p})(x_0 - y_0)) - i \sin(\omega(\mathbf{p})(x_0 - y_0)) \right)
\end{aligned}
\tag{4.13}
$$

For high temperature or $\hbar \rightarrow 0$ the denominator of the first term ($\mu = 0$) is ω^{-2}, which gives a correlation function in classical field theory (the second term is vanishing in the limit $\hbar \rightarrow 0$). Hence, in the limit $\hbar \rightarrow 0$

$$
< \phi(x)\phi(y) >_\beta = \beta^{-1}(2\pi)^{-3} \int \frac{d\mathbf{p}}{\omega(\mathbf{p})^2} \cos(\omega(\mathbf{p})(x_0 - y_0)) \cos(\mathbf{p}(\mathbf{x} - \mathbf{y})). \tag{4.14}
$$

The correlation function (4.14) is equal to

$$
\int d\phi \exp(-\beta H)\phi(x)\phi(y),
$$

where $H(\phi)$ is the classical Hamiltonian (2.25).

It can easily be seen that

$$
\begin{aligned}
&< T(\phi(x)\phi(y)) >_\beta = i\Delta_F(x - y) + 2\hbar(2\pi)^{-3} \int \frac{d\mathbf{p}}{2\omega(\mathbf{p})} \\
&\times \left(\exp(\beta(\omega(\mathbf{p}) - \mu)) - 1 \right)^{-1} \cos(\omega(\mathbf{p})(x_0 - y_0)) \cos(\mathbf{p}(\mathbf{x} - \mathbf{y})).
\end{aligned}
\tag{4.15}
$$

We can calculate all correlation functions from

$$
< T(\exp(\phi(f))) >_\beta = \exp\left(\frac{1}{2} \int dx dy f(x) f(y) < T(\phi(x)\phi(y)) >_\beta \right). \tag{4.16}
$$

4.2   Scalar Free Field on a Globally Hyperbolic Manifold

In this section we discuss the quantization of a scalar field on a globally hyperbolic manifold [147] . On such a manifold we can choose coordinates which transform the metric to the form [31]

$$
ds^2 = g_{00}dx^0 dx^0 - g_{jk}dx^j dx^k \equiv g_{\mu\nu}dx^\mu dx^\nu. \tag{4.17}
$$

The Lagrangian of the free field is

$$
\mathcal{L} = \frac{1}{2}\sqrt{g}g^{\mu\nu}\partial_\mu\phi\partial_\nu\phi - \frac{m^2}{2}\sqrt{g}\phi^2, \tag{4.18}
$$

where $g = |\det[g_{\mu\nu}]|$. The Klein-Gordon equation resulting from the Lagrangian has the form

$$g^{-\frac{1}{2}}\partial_\mu g^{\frac{1}{2}} g^{\mu\nu} \partial_\nu \phi + m^2 \phi = 0. \tag{4.19}$$

The canonical momentum is

$$\Pi = \frac{\partial \mathcal{L}}{\partial \partial_0 \phi} = g^{00} \sqrt{g} \partial_0 \phi. \tag{4.20}$$

The Hamiltonian density reads

$$T_{00}(x) \equiv H(x) = \Pi \partial_0 \phi - L = \frac{1}{2} g_{00} \frac{1}{\sqrt{g}} \Pi^2 + \frac{1}{2} \sqrt{g} g^{jk} \partial_j \phi \partial_k \phi + \frac{m^2}{2} \sqrt{g} \phi^2. \tag{4.21}$$

We restrict ourselves in this section to some special cases of the metric (4.17) (for some generalizations see Chap. 8). Assume that $g^{\mu\nu}$ is time-independent (the Schwarzschild metric in static coordinates is of this form). Define the operator [99]

$$K^2 = -g_{00} \frac{1}{\sqrt{g}} \partial_j g^{jk} \sqrt{g} \partial_k + g_{00} m^2 \tag{4.22}$$

which is symmetric with respect to the measure

$$dv = g^{00} \sqrt{g} d\mathbf{x}. \tag{4.23}$$

The Klein-Gordon equation (4.19) can be expressed in the form [99]

$$\partial_t^2 \phi = -K^2 \phi. \tag{4.24}$$

This is a general expression of the wave equation. We write this equation in the first order form

$$\frac{dX}{dt} \equiv MX = \begin{bmatrix} 0 & 1 \\ -K^2 & 0 \end{bmatrix} X, \tag{4.25}$$

where X is a column, $X^T = (\phi, P)$. The solution is expressed by the initial condition X_0 as $X_t = \exp(Mt)X_0$ as discussed in [203]. In the Minkowski space $P = \Pi$ we can express explicitly (ϕ, P) at time t [112] by time zero Cauchy data (Δ is the commutator function (2.42))

$$\phi(x) = i \int \partial_0 \Delta(t, \mathbf{x} - \mathbf{y}) \phi(\mathbf{y}) d\mathbf{y} + \int \Delta(t, \mathbf{x} - \mathbf{y}) P(\mathbf{y}) d\mathbf{y},$$

$$P(x) = i \int \partial_0 \Delta(x_0, \mathbf{x} - \mathbf{y}) P(\mathbf{y}) d\mathbf{y} + \int \partial_0^2 \Delta(x_0, \mathbf{x} - \mathbf{y}) \phi(\mathbf{y}) d\mathbf{y},$$

as can be shown by direct differentiation. (ϕ, P) satisfy Hamilton equations of motion. It remains to check that the initial conditions are satisfied. On a globally hyperbolic manifold there exists a solution of the Klein-Gordon initial value problem in the form [54, 239]

$$\phi_t(\phi, P) = K_1\phi + K_2 P, \tag{4.26}$$

where K_1 and K_2 are some kernels in the space of functions defined on the zero-time section of the manifold (the integration in (4.26) is with respect to the Riemannian measure on the time-zero section).

For a quantum theory we must define fields which solve Eq. (4.24) in Hilbert space. As discussed in Sect. 2.5 in the Minkowski space-time the mathematically well-defined Hilbert space is $L^2(d\mu)$, where the Gaussian measure μ can be defined using the ground state ψ^g as $d\mu = |\psi^g|^2 d\phi$. In the heuristic Hilbert space $L^2(d\phi)$ we have

$$\Pi(\mathbf{x}) = -i\hbar \frac{\delta}{\delta\phi(\mathbf{x})} \tag{4.27}$$

which we insert in Eq. (4.21) in order to define the heuristic Hamiltonian. We show that if the metric is time-independent then

$$\psi^g = Z^{-1} \exp\left(-\frac{1}{2\hbar}(\phi, K\phi)\right) \tag{4.28}$$

is the ground state solution for the Hamiltonian H (4.21). Z is a normalization factor such that $\int d\phi |\psi_g|^2 = 1$. In Eq. (4.28)

$$(\phi, \phi') = \int d\mathbf{x}\, g^{00}\sqrt{g}\,\phi(\mathbf{x})\phi'(\mathbf{x}). \tag{4.29}$$

In order to prove that $H\psi^g = E_g\psi = \hbar \int d\mathbf{x}\, K(\mathbf{x}, \mathbf{x})\psi^g$ where E_g is the ground state energy, we need to calculate

$$\begin{aligned}
&\frac{\hbar^2}{2}\int d\mathbf{x}\, g_{00}g^{-\frac{1}{2}}\frac{\delta^2}{\delta\phi(\mathbf{x})^2}\exp\left(-\frac{1}{2\hbar}\int d\mathbf{x}\, g^{00}g^{\frac{1}{2}}\phi(\mathbf{x})(K\phi)(\mathbf{x})\right)\\
&= \left(\int d\mathbf{x}\left(\tfrac{1}{2}\sqrt{g}g^{jk}\partial_j\phi\partial_k\phi + \tfrac{m^2}{2}\sqrt{g}\phi^2\right) - \tfrac{1}{2}\hbar\int d\mathbf{x}\, K(\mathbf{x},\mathbf{x})\right)\psi^g.
\end{aligned} \tag{4.30}$$

In Eq. (4.30) we used the fact that K^2, hence also K, are symmetric operators in the scalar product (4.29). Then, the term calculated in Eq. (4.30) (corresponding to the first term on the rhs of $H\psi^g$ in Eq. (4.21)) cancels with second term on the rhs of $H\psi^g$ Eq. (4.21). The last term in Eq. (4.30) gives the vacuum energy. We construct the Hilbert space according to Sect. 2.5 as $L^2(d\mu)$ with $d\mu = |\psi^g|^2 d\phi$, where the mathematical definition of μ is through its Fourier transform

$$\int d\mu(\phi) \exp(i(\phi, f)) = \exp\left(-\frac{\hbar}{4}(f, K^{-1}f)\right), \qquad (4.31)$$

where the scalar product (ϕ, f) is defined in Eq. (4.28). Now, the physical Hamiltonian is defined as $H_{ph} = \psi_g^{-1} H \psi_g$. Its ground is 1. The canonical momentum is $(\psi^g)^{-1}\left(-i\hbar\frac{\delta}{\delta\phi}\right)\psi^g$. If we calculate the fields $\phi_t(\phi, P)$ from Eq. (4.25) then we can compute the correlation functions as

$$(\psi^g, \phi(\mathbf{x})\phi_t(\mathbf{x}')\psi^g) = \int d\mu(\phi)\phi(\mathbf{x})\phi_t(\mathbf{x}', \phi, P)1, \qquad (4.32)$$

where on the rhs of Eq. (4.32) we have the action of the field ϕ_t on the vacuum 1 in $L^2(d\mu)$. $\phi_t(\mathbf{x}', \phi, P)$ is the solution of Eq. (4.25) expressed by the initial conditions ϕ and P. This solution should be the same as the one obtained from

$$\phi_t(\mathbf{x}) = \exp\left(\frac{i}{\hbar}H_{ph}t\right)\phi(\mathbf{x})\exp\left(-\frac{i}{\hbar}H_{ph}t\right)$$

with correlations calculated as $(1, \phi(\mathbf{x})\phi_t(\mathbf{x}')1)$. For more about quantization on a curved manifold see [35, 66, 239] and Sects. 8.6–8.9.

Another way to construct the Hilbert space is based on the Fock method [99]. We consider eigenfunctions of K

$$Kf_j = \omega_j f_j.$$

We expand quantum fields in eigenfunctions of K

$$\phi = \sum_j a_j f_j \exp(-i\omega_j t) + (a_j f_j)^+ \exp(i\omega_j t). \qquad (4.33)$$

We can construct the Hilbert space as the Fock space (Sect. 2.3) generated by the creation operators from the vacuum $|0>$ defined by $a|0>= 0$. It is straightforward to express the Feynman propagator by the operator K. We repeat the calculations at Eq. (2.41) with $\omega(p) \to \omega_j$ using the completeness of the basis functions f_j in Eq. (4.33). The result is

$$\Delta^{(+)}(x, x') = \left((2K)^{-1}\exp(-iK(t-t'))\right)(\mathbf{x}, \mathbf{x}'), \qquad (4.34)$$

$$i\Delta_F(x, x') = \left((2K)^{-1}\exp(-iK|t-t'|)\right)(\mathbf{x}, \mathbf{x}'). \qquad (4.35)$$

If $g_{\mu\nu}$ is time-independent then there is no difficulty with the transition from imaginary time to real time and back. We can also define fields at finite temperature replacing $\omega(\mathbf{p})$ in Eq. (4.9)with ω_j and f_p in Eq. (4.10) by f_j from Eq. (4.33). This enforces the corresponding changes in Eq. (4.13).

From the real time (4.32)–(4.34) we can derive QFT at imaginary time (Euclidean version) by means of an analytic continuation. It is instructive to consider the imaginary time version of the free QFT on the static manifold in the Osterwalder-Schrader formulation of Sect. 3.10.

The analytic continuation of the propagator (4.35) is

$$
\begin{aligned}
G_K(x, x') &= \Big((2K)^{-1} \exp(-K|t - t'|) \Big)(\mathbf{x}, \mathbf{x}') \\
&= (2\pi)^{-1} \int dp_0 \exp(i p^\mu (x^\mu - x'^\mu))(K^2 + p_0^2)^{-1}.
\end{aligned}
\tag{4.36}
$$

We consider the generating functional

$$
Z[f] = \exp\left(-\frac{1}{2} f G_K f \right).
\tag{4.37}
$$

It satisfies the Minlos-Sazonov condition (1.14) for an existence of the Gaussian measure. For the proof we can repeat the argument (1.16) (applied now to G_K of Eq. (4.36)) leading to the inequality (1.14). Hence, (4.37) defines Euclidean fields. We are interested whether the Osterwalder-Schrader positivity of Sect. 3.10 is satisfied. We define $\vartheta t = -t$ then for $t \geq 0$ and $t' \geq 0$

$$
G_K(x, \vartheta x') = \Big((2K)^{-1} \exp(-K(t + t')) \Big)(\mathbf{x}, \mathbf{x}').
\tag{4.38}
$$

It can be seen that Eq. (4.38) defines a positive definite kernel on the space of functions with a positive time as

$$
\begin{aligned}
&\int dt dt' d\mathbf{x} d\mathbf{x}' f^*(t, \mathbf{x}) f(t', \mathbf{x}') G_K(x, x') \\
&= \int dt dt' d\mathbf{x} \Big((2K)^{-\frac{1}{2}} \exp(-Kt) f \Big)^* (\mathbf{x}) \Big((2K)^{-\frac{1}{2}} \exp(-Kt') f \Big)(\mathbf{x}) \geq 0
\end{aligned}
$$

With the representation (4.38) we can repeat the proof of OS positivity (Eq. (3.89) based on Eq. (1.16))'of Sect. 3.10 which exploited a similar representation of the Euclidean two-point function of the free field). It follows from Sect. 3.10 that the Gaussian measure with the covariance (4.36) defines a random field whose generating functional $Z[f]$ (4.37) satisfies the OS positivity (3.82). For real time we have the Wightman reconstruction theorem of quantum fields from vacuum correlation functions (4.34) [228]. At imaginary time (4.37) after the proof of OS positivity the reconstruction theorem of Osterwalder and Schrader [191] allows to construct the quantum fields in a Hilbert space from (commutative) Euclidean fields. For more results on Euclidean QFT with OS positivity on a manifold see [61, 65, 136, 164]).

As a next example of a model quantized in the real time we consider a spatially flat manifold with an expanding metric such that $g_{00} = 1$ and

$$
g_{ij} = a(t)^2 \delta_{ij}
$$

(half of de Sitter space-time is of this type, see Sect. 8.9). We select a complete set of solutions $F_j(t, \mathbf{p})$ of the Klein-Gordon equation (in the momentum space) with a positive energy defined by the asymptotic behavior of F_j at large time

$$F_j(t, \mathbf{p}) \to f_j(\mathbf{p}) \exp(-i\omega_j t). \tag{4.39}$$

The quantum field is defined by [35]

$$\phi = \sum_j a_j F_j + (a_j F_j)^+ \tag{4.40}$$

in the Fock space equipped with the vacuum $|0>$ defined by $a_j|0>= 0$. There remains the problem of the non-uniqueness of this construction following from the fact that the choice of the basis functions F_j is not unique. We still discuss this model in Sects. 8.8–8.9.

Instead of describing the free field theory in the Hilbert space $L^2(d\mu)$ where the measure μ is a realization (Sect. 1.3) of the heuristic $d\mu = |\psi^g|^2 d\phi$ we could work in the Wigner phase space [242]. For a given density matrix ρ (in the infinite dimensional field functional space [82]) the Wigner density function is defined by the heuristic formula

$$p(\phi, \Pi) = \int d(\delta\phi) \exp\left(-\frac{i}{\hbar} \int d\mathbf{x} \Pi(\mathbf{x})\delta\phi(\mathbf{x})\right) \rho\left(\phi + \frac{1}{2}\delta\phi, \phi - \frac{1}{2}\delta\phi\right) \tag{4.41}$$

(we do not normalize p here). We can define the probability measure $d\mu^W(\phi, \Pi) = N d\phi d\Pi p(\phi, \Pi)$ (where N is the normalization) by its Fourier transform

$$Z[f, h] = \int d\mu^W(\phi, \Pi) \exp\left(i(\phi, f) + i \int d\mathbf{x} \Pi(\mathbf{x})h(\mathbf{x})\right). \tag{4.42}$$

If p is Gaussian then Eq. (4.42) is mathematically well-defined in the sense of Sect. 1.3. For a pure state

$$\rho(\phi, \phi') = \psi^*(\phi)\psi(\phi'). \tag{4.43}$$

Then, the Wigner measure corresponding to the ground state (4.28) is

$$d\mu^W(\phi, \Pi) = d\phi d\Pi \exp \int d(\delta\phi) \exp\left(-\frac{1}{2\hbar}((\phi - \tfrac{1}{2}\delta\phi), K(\phi - \tfrac{1}{2}\delta\phi))\right)$$
$$\times \exp\left(-\frac{1}{2\hbar}((\phi + \tfrac{1}{2}\delta\phi), K(\phi + \tfrac{1}{2}\delta\phi))\right) \exp(-\tfrac{i}{\hbar} \int d\mathbf{x} \Pi(\mathbf{x})\delta\phi(\mathbf{x}))$$
$$= d\phi d\Pi \exp\left(-\frac{1}{\hbar}(\phi, K\phi) - \hbar(\Pi, K^{-1}\Pi)_{-1}\right).$$

where $(\phi, K\phi)$ is defined by the scalar product in the Hilbert space $L^2(dv)$ with dv of Eq. (4.23) and

$$(\Pi, K^{-1}\Pi)_{-1} = \int d\mathbf{x} g_{00} g^{-\frac{1}{2}} \Pi(\mathbf{x})(K^{-1}\Pi)(\mathbf{x}).$$

Calculating the Gaussian integrals with respect to $d\mu^W(\phi, \Pi)$ in Eqs. (4.41)–(4.42) we obtain the positively semi-definite functional of the Minlos-Sazonov theorem (Sect. 1.3) as

$$Z[f, h] = \exp\left(-\frac{\hbar}{4}(f, K^{-1}f)\right) \exp\left(-\frac{1}{4\hbar}(h, Kh)\right) \qquad (4.44)$$

From Eqs. (4.42) and (4.44) we obtain

$$< \phi(f)^2 >< \Pi(h)^2 >= \frac{\hbar^2}{4}(f, K^{-1}f)(h, Kh) \geq \frac{\hbar^2}{4}(f, h)^2$$

in agreement with Heisenberg uncertainty relations.

From Eqs. (4.41)–(4.42) we can conclude that the expectation values with respect to the Wigner measure $d\mu^W$ are equal to the expectation values in the ground state

$$\int d\mu^W \mathcal{F}(\phi) = (\psi^g, \mathcal{F}(\phi)\psi^g) \qquad (4.45)$$

$$\int d\mu^W \mathcal{G}(\Pi) = (\psi^g, \mathcal{G}(\Pi)\psi^g) \qquad (4.46)$$

The equalities (4.45)–(4.46) can still hold true for an operator $\mathcal{F}(\phi, \Pi)$ if the corresponding function $\mathcal{F}(\phi, \Pi)$ is constructed following the Wigner-Weyl transform [89]. The Wigner function (4.41) is constructed in such a way that if ψ is the solution of the Schrödinger equation $i\hbar\partial_t\psi = H\psi$ with the Hamiltonian (4.21) then $p(\phi, \Pi)$ satisfies the classical Liouville evolution equation for the probability density on the phase space

$$\partial_t p(\phi, \Pi) = \int d\mathbf{x} g^{00}\sqrt{g}\Pi\frac{\delta p}{\delta\phi(\mathbf{x})} + \int d\mathbf{x}(\partial_j\sqrt{g}g^{jk}\partial_k\phi - m^2\sqrt{g}\phi)\frac{\delta p}{\delta\Pi(\mathbf{x})} \qquad (4.47)$$

The generating functional (4.44) can be used to calculate all correlation functions in the phase space. Such correlation functions can be related to operator expectation values in the ground state by means of the Wigner-Weyl transform [89]. We shall still discuss the quantization of fields on manifolds with a time dependent metric in the framework of functional integration in Chap. 8. When the metric depends on time then we do not expect any ground state for the Hamiltonian (4.21). However, there is a Gaussian time-dependent solution of the Schrödinger equation which defines a measure on the phase space. This measure can be considered as a state on the algebra of observables in the sense of the algebraic quantum field theory [132, 239].

4.3 Exercises

4.1 Prove Eqs. (4.2)–(4.3).

4.2 Discuss an expansion of Eq. (4.13) for $\beta \to 0$ and $\beta \to \infty$.

4.3 Prove the formula (4.16) for the generating functional of thermal correlation functions.

4.4 Show that ψ_g (4.28) is the ground state of the Hamiltonian (4.21) as indicated in Eq. (4.30).

4.5 Establish the formula (4.22) for the solution of the Cauchy problem for the Klein-Gordon equation on the Minkowski space-time.

4.6 Calculate $< 0|T(\phi(x)\phi(y))|0 >$ for the Fock space representation (4.33).

4.7 In an expanding metric express $< 0|T(\phi(x)\phi(y))|0 >$ in terms of the classical solutions F_j of Eq. (4.39).

4.8 Show that the Wigner function of Eq. (4.41) satisfies the Liouville equation (4.47) for the classical probability density.

4.9 Calculate the exponential of the matrix M in Eq. (4.25). Show that the ϕ-component of $\exp(tM)$ leads to the solution

$$\phi_t(\phi, \Pi) = \cos(Kt)\phi + K^{-1}\sin(Kt)\Pi$$

Chapter 5
The Functional Integral

Abstract We formulate the functional integration for Hamiltonian evolution first in quantum mechanics which can be considered as quantum field theory in one space-time dimension. Feynman formulated his path integral integral on the physical basis of an interference of short time amplitudes. The composition of short time amplitudes has a mathematical version of the Trotter product formula. If the potentials and wave functions have an analytic continuation in the complex plane then the Feynman integral can be expressed by a well-defined Wiener measure. We show that if the perturbation series in the potential is convergent then the sum of the series is equal to the Feynman integral. The Hamiltonian with an electromagnetic field requires the path integration with a stochastic integral which will be used also in subsequent chapters. In this chapter we explain the calculus with the stochastic integrals. We develop methods to approach non-perturbative estimates on quantum expectation values in real and imaginary time.

We formulate the functional integration for Hamiltonian evolution first in quantum mechanics which according to the dimensional regularization of Sect. 3.8 may be considered as quantum field theory in one space-time dimension or in zero spatial dimension. From the beginning we have emphasized that QFT is the quantum mechanics with an infinite number of degrees of freedom. The formulas which we have derived for QFT in earlier sections apply to quantum mechanics as well (and are non-trivial there). Feynman formulated his path integral integral [87] on the physical basis of an interference of short time amplitudes. The composition of short time amplitudes has a mathematical version of the Trotter formula (Sect. 5.1). If the potentials and wave functions have an analytic continuation in the complex plane then the Feynman integral can be expressed by a well-defined Wiener measure (Sect. 5.3). We show that for some potentials and wave functions the Wiener integral over analytically continued paths gives the same result as the Feynman integral. We prove that if the perturbation series in the potential is convergent then the sum of the series is equal to the Feynman integral. The Wiener measure gives a probability distribution of the Brownian motion: a diffusion process of a free particle. The Hamiltonian with an electromagnetic field requires the path integration with a stochastic integral (Sect. 5.4) which will be used also in subsequent chapters. The original path integral is formulated in terms of the Gaussian integral describing the paths of the free particle. If we wish to integrate over the paths of an interacting particle we need to transform the paths by means of a stochastic equation. Further on we shall mainly use a transformation from a stochastic path of a free particle to the oscillator's paths (applied in QFT). The basic theory of the stochastic equations is discussed in Sect. 5.5.

As an illustration of the relation between the results of QFT discussed in Chap. 4 and quantum mechanics we express the annihilation operator by the position and momentum operators

$$a = 2^{-\frac{1}{2}} \left(\sqrt{\omega} x + \frac{i}{\sqrt{\omega}} p \right). \tag{5.1}$$

Then,

$$\sqrt{2\omega} x = a + a^+. \tag{5.2}$$

We have $H_0 = \omega a^+ a$. The oscillator ground state $|0>$ is defined by $a|0>= 0$. The Heisenberg evolution reads

$$x(t) = \exp(i H_0 t) x \exp(-i H_0 t) = (2\omega^2)^{-\frac{1}{2}} (\exp(-i\omega t)a + \exp(i\omega t)a^+). \tag{5.3}$$

Equation (5.3) looks like the expansion (2.22) of the field in creation-annihilation operators. We may introduce an interaction $V(x)$ and apply the interaction picture in order to compute the correlation functions of the position operator in the true ground state Φ_0 defined by

$$(H_0 + V)\Phi_0 = 0. \tag{5.4}$$

The correlation functions in the ground state Φ_0 could be expressed as the correlation functions in the ground state of the harmonic oscillator via the Gell-Mann-Low formula

$$\begin{aligned} &< \Phi_0 | T(x(t_1)......x(t_n)) | \Phi_0 > \\ &= Z^{-1} < 0 | T\left(x(t_1)......x(t_n) \exp(-i \int_{-\infty}^{\infty} ds\, V(x(s))) \right) | 0 >, \end{aligned} \tag{5.5}$$

where Z is the Gell-Mann-Low normalization factor.

The formula (5.5) is non-trivial and useful for quantum mechanics of the anharmonic oscillator. The indefinite integral $\int_{-\infty}^{\infty} ds\, V$ needs a definition as a limit of an oscillatory integral which is difficult to establish. However, the imaginary time version of the Gell-Mann-Low formula can be proved (see [58]). In this chapter we develop methods to approach non-perturbative estimates on quantum expectation values.

5.1 Trotter Product Formula and The Feynman Integral

In this section we derive the Feynman integral for quantum mechanics. As discussed at the beginning of this chapter quantum field theory can be considered as quantum mechanics with an infinite number of degrees of freedom. A heuristic generalization to quantum field theory is straightforward. The Feynman formula may be considered as a new method of quantization applied to theories where the canonical formalism (because of redundant degrees of freedom) is not obvious. It is our aim in this section to show how the Feynman integral follows from the Hamiltonian quantum mechanics. The derivation comes from the Trotter formula for a semigroup of operators [235]

$$\exp(-t(A + B)) = \lim_{n \to \infty} \left(\exp\left(-\frac{t}{n}A \right) \exp\left(-\frac{t}{n} \right) \right)^n. \tag{5.6}$$

If $[A, B] = 0$ then Eq. (5.6) is an identity even without the limit. We can prove Eq. (5.6) for matrices showing that

$$\exp\left(-\frac{t}{n}(A + B) \right) = \exp\left(-\frac{t}{n}A \right) \exp\left(-\frac{t}{n}B \right) + O(n^{-2}), \tag{5.7}$$

because $[\frac{t}{n}A, \frac{t}{n}B] \simeq \frac{t^2}{n^2}$. Equation (5.7) can be shown using Campbell-Baker formula or expanding both sides in $\frac{1}{n}$. It says that for a small time we can ignore non-commutativity and approximate in Eq. (5.7) the lhs by the first term on the rhs. For unbounded operators we apply (5.6) to states in a Hilbert space. The convergence is in the Hilbert space norm [203] (Theorem VIII. 30). An estimate of the error for finite n in Eq. (5.7) is important for numerical simulations (it depends on the state [47]). We apply the Trotter formula to the Schrödinger evolution. Then,

$$A = -i\frac{\hbar}{2m}\Delta, \tag{5.8}$$

$$B = \frac{i}{\hbar}V \tag{5.9}$$

corresponding to the Hamiltonian evolution with $H = H_0 + V$ where H_0 is the Hamiltonian for a free particle.

Feynman integral results when we rewrite the Trotter product formula (5.6) in terms of kernels (see Sect. 3.4). We need the kernel (for a simplicity of notation we restrict ourselves to 1-dimension)

$$\exp(-tA)(x, y) = \left(\frac{2\pi i \hbar t}{m} \right)^{-\frac{1}{2}} \exp\left(\frac{im(x - y)^2}{2t\hbar} \right). \tag{5.10}$$

The sign of the square root in Eq. (5.10) is fixed by the requirement that

$\lim_{t \to 0} \exp(-tA)(x, y) = \delta(x - y)$. Equation (5.10) can be proved when we notice that this kernel is a Fourier transform of the evolution operator in the momentum space which is

$$\exp\left(-\frac{ip^2}{2m\hbar}t\right). \tag{5.11}$$

The Fourier transform can be calculated using the formula for Gaussian integration (Sect. 1.3) which we apply many times in these lectures (Eq. (1.13))

$$\int dx \exp(-\tfrac{a}{2}x^2) \exp(ux) \left(\int dx \exp(-\tfrac{a}{2}x^2)\right)^{-1} \tag{5.12}$$
$$= \exp(\tfrac{u^2}{2a}).$$

In order to show the formula (5.12) it is sufficient to introduce a new variable $y = x - \frac{u}{a}$. Then, the nominator cancels with the denominator and what remains is the rhs of Eq. (5.12).

We still need the kernel

$$\exp(-tB)(x, y) = \delta(x - y) \exp\left(-\frac{i}{\hbar}tV(x)\right) \tag{5.13}$$

We have shown in Sect. 3.4 that the kernel of the product of operators is expressed by a product of kernels. So, a typical term in Eq. (5.6) has the form

$$\left(\exp(-\tfrac{t}{n}A) \exp(-\tfrac{t}{n}B)\right)(x_k, x_{k+1})$$
$$= \int dz (\exp(-\tfrac{t}{n}A))(x_k, z) \exp(-\tfrac{t}{n}B)(z, x_{k+1}) \tag{5.14}$$
$$= \left(\frac{2\pi i \hbar t}{nm}\right)^{-\frac{1}{2}} \exp\left(\frac{inm(x_k - x_{k+1})^2}{2t\hbar}\right) \exp(-\tfrac{it}{n\hbar}V(x_{k+1})).$$

Clearly, in Eq. (5.6) we have a product of such terms and an integration $dx_1......dx_n$ over the points x_k.

How to interpret Eq. (5.6)? Note that the expression in the exponential is a discrete approximation to the Riemannian integral

$$\frac{i}{\hbar}\int_{\frac{tk}{n}}^{\frac{t(k+1)}{n}} ds \left(\frac{m}{2}\left(\frac{dx}{ds}\right)^2 - V(x(s))\right) = \frac{i}{\hbar}\int_{\frac{tk}{n}}^{\frac{t(k+1)}{n}} \mathcal{L}(x(s))ds,$$

where $\mathcal{L}$ is the particle Lagrangian $\mathcal{L} = \frac{m}{2}(\frac{dx}{ds})^2 - V(x)$. In order to obtain the exponential of Eq. (5.14) we approximate

$$\frac{dx}{ds} \simeq \left(x\left(s + \frac{t}{n}\right) - x(s)\right)\left(\frac{t}{n}\right)^{-1} \tag{5.15}$$

and

$$\int\limits_{\frac{tk}{n}}^{\frac{t(k+1)}{n}} ds\, V(x(s)) \simeq V\left(x((k+1)\frac{t}{n})\right)\frac{t}{n}.$$

On the rhs of Eq. (5.6) we have an exponential of the sum of the terms in the exponentials (5.14) and integrals over the intermediate points x_k. These are the points on the trajectory $x(s)$. Hence, we may interpret the integration over x_k as an integration over the trajectory $x(s)$.

We have reached the formula

$$\left(\exp\left(-\frac{i}{\hbar}Ht\right)\right)(x, y) = \int\limits_{(x,y)} \mathcal{D}x \exp\left(\frac{i}{\hbar}\int\limits_0^t ds\mathcal{L}(x(s))\right) \qquad (5.16)$$

(x, y) means that we integrate over paths ($\mathcal{D}x$-integral) which start in x and end in y.

The evolution of the wave function ψ results from an integration of the kernel (5.16)

$$\left(\exp\left(-\frac{i}{\hbar}Ht\right)\psi\right)(x) = \int \left(\exp\left(-\frac{i}{\hbar}Ht\right)\right)(x, y)\psi(y)dy.$$

Hence, in terms of paths

$$\left(\exp\left(-\frac{i}{\hbar}Ht\right)\psi\right)(x) = \int\limits_{x(0)=x} \mathcal{D}x \exp\left(\frac{i}{\hbar}\int_0^t ds\mathcal{L}(x(s))\right)\psi(x(t)), \qquad (5.17)$$

where now the integral is over all paths starting from x. The virtue of the Feynman formula is that for any classical model with the Lagrangian $\mathcal{L}$ it gives a prescription for the quantum evolution [87, 170].

Let us still write the imaginary time version of this formula (called Feynman-Kac formula)

$$\left(\exp\left(-\frac{1}{\hbar}Ht\right)\right)(x, y) = \int\limits_{(x,y)} \mathcal{D}x \exp\left(-\frac{1}{\hbar}\int\limits_0^t ds\mathcal{L}_E(x(s))\right), \qquad (5.18)$$

where

$$\mathcal{L}_E = \frac{m}{2}\left(\frac{dx}{ds}\right)^2 + V(x). \qquad (5.19)$$

In order to prove (5.18) we represent it in terms of kernels of operators

$$\left(\exp\left(-\frac{t}{\hbar n}H_0\right)\exp\left(-\frac{t}{\hbar n}V\right)\right)^n.$$

The benefit of the imaginary time version is that we avoid the oscillatory integrals which are difficult for numerical approximations.

5.2   Evolution for Time-Dependent Hamiltonians

For time-dependent Hamiltonians of the form

$$H(t) = -\frac{\hbar^2}{2m}\Delta + V(t, \mathbf{x}) = H_0 + V(t, \mathbf{x}) \tag{5.20}$$

the solution of the Schrödinger equation can be expressed by the time-ordered exponential [70] (similarly as in the interaction picture in Sect. 3.1)

$$\hat{U}(t, s) = T\left(\exp\left(-\frac{i}{\hbar}\int_s^t H(t')dt'\right)\right), \tag{5.21}$$

i.e., the solution with the initial condition ψ at the initial time s is $\hat{U}(t, s)\psi$.

For $\hat{U}(t, t_0)$ we have an analog of the Trotter product formula

$$\hat{U}(t, t_0) = \lim_{n\to\infty} \prod_k \exp\left(-\frac{i(t - t_0)}{n\hbar}V\left(t_0 + \frac{k(t - t_0)}{n}\right)\right)\exp\left(-\frac{i(t - t_0)}{n\hbar}H_0\right), \tag{5.22}$$

where $0 \leq k \leq n$.

Beginning from the rhs of $\hat{U}(t, t_0)\psi$ we obtain

$$(\hat{U}(t, t_0)\psi)(x)$$
$$= \int Dx \int \exp\left(\frac{i}{\hbar}\int_{t_0}^t ds \frac{1}{2}(\frac{dx}{ds})^2 - \frac{i}{\hbar}\int_{t_0}^t ds V(s, x(s))\right)\psi(x(0)). \tag{5.23}$$

In order to relate it with the formula (5.17) we change the paths as $x(s) \to x(t - s)$ changing also the integration variable $s \to t - s$. In such a case the path integral is transformed to

$$(\hat{U}(t, t_0)\psi)(x)$$
$$= \int Dx \int \exp\left(\frac{i}{\hbar}\int_{t_0}^t ds \frac{1}{2}(\frac{dx}{ds})^2 - \frac{i}{\hbar}\int_{t_0}^t ds V(t - s, x(s))\right)\psi(x(t)). \tag{5.24}$$

Another way to treat the time-dependent Hamiltonians of particles in R^d is to enlarge the space of functions to $L^2(R^{d+1})$ and consider the "Schrödinger operator" K in $L^2(R^{d+1})$ [155, 203] (with $K_0 = -i\hbar\partial_t$)

$$K = -i\hbar\partial_t + H(t) \equiv K_0 + H(t). \tag{5.25}$$

We consider the equation

$$i\hbar\partial_\tau f = Kf. \tag{5.26}$$

Let

$$U(\tau) = \exp\left(-\frac{i}{\hbar}\tau K\right). \tag{5.27}$$

Denote $T(\tau) = \exp(-\frac{i\tau}{\hbar}K_0) = \exp(-\tau\partial_t)$. We show that

$$U(\tau) = \hat{U}(t, t - \tau)T(\tau). \tag{5.28}$$

To prove (5.28) we must show that

$$U(\tau)f(t, x) = \hat{U}(t, t - \tau)f(t - \tau, \mathbf{x}). \tag{5.29}$$

We differentiate both sides of Eq. (5.29) over τ and obtain Eq. (5.26).

The rhs of Eq. (5.29) can be expressed as the Feynman formula for $\hat{U}(t, t - \tau)$

$$T\left(\exp(-\frac{i}{\hbar}\int_{t-\tau}^{t} H(s)ds)\right)\psi)(x)$$
$$= \int \mathcal{D}x \exp\left(\frac{i}{\hbar}\int_{t-\tau}^{t} ds\frac{1}{2}(\frac{dx}{ds})^2 - \frac{i}{\hbar}\int_{t-\tau}^{t} dsV(t-s, x(s))\right)\psi(x(t))$$

leading to the formula (5.24).

5.3   The Wiener Integral and Feynman-Wiener Integral

The rigorous version of the Feynman-Kac formula (5.16)–(5.17) is expressed in the imaginary time as in Eq. (5.18) (we set in this section $\hbar = m = 1$)

$$((\exp(-tH)\psi)(\mathbf{x})$$
$$= \int dW(\mathbf{w})\exp\left(-\int_0^t V(\mathbf{x} + \mathbf{w}(s))ds\right)\psi(\mathbf{x} + \mathbf{w}(t)), \tag{5.30}$$

where dW is the Wiener measure describing the Gaussian process $\mathbf{w}(s) \in R^3$ ($s \geq 0$, $\mathbf{w}(0) = 0$) with mean zero and the covariance

$$E[w^j(t)w^k(s)] = \delta^{jk}min(t, s). \tag{5.31}$$

The Wiener integral can be defined by its value on the "cylinder functions" [104, 109]

$$\int dW(w)F(\mathbf{w}(s_1), \mathbf{w}(s_2), \ldots, \mathbf{w}(s_n)) = \int d\mathbf{x}_1 \ldots d\mathbf{x}_n F(\mathbf{x}_1, \ldots, \mathbf{x}_n)$$
$$\times p(s_1, \mathbf{x}_1)p(s_2 - s_1, \mathbf{x}_2 - \mathbf{x}_1)\ldots p(s_n - s_{n-1}, \mathbf{x}_n - \mathbf{x}_{n-1}), \tag{5.32}$$

where

$$p(s, \mathbf{x} - \mathbf{y}) = (2\pi s)^{-\frac{3}{2}} \exp\left(-\frac{(\mathbf{x} - \mathbf{y})^2}{2s}\right). \qquad (5.33)$$

The kernel of $\exp(-tH)$ can be written in the form

$$(\exp(-tH))(\mathbf{x}, \mathbf{y}) = \int dW(\mathbf{w})\delta(\mathbf{w}(t) - \mathbf{y}) \exp\left(-\int_0^t V(\mathbf{x} + \mathbf{w}(s))ds\right). \quad (5.34)$$

We can use (5.34) to express the kernel of the operator $\exp(-tH)$ as [109, 224]

$$\exp(-tH)(\mathbf{x}, \mathbf{y}) \equiv \int dW_{(\mathbf{x},\mathbf{y})}^t(\mathbf{q}) \exp(-\int_0^t V(\mathbf{q}(s))ds) = (2\pi t)^{-\frac{3}{2}}$$
$$\times \exp(-\frac{(\mathbf{x}-\mathbf{y})^2}{2t}) \int d\nu(\mathbf{r}) \exp\left(-\int_0^t V(\mathbf{x} + \frac{s}{t}(\mathbf{y} - \mathbf{x}) + \sqrt{t}\mathbf{r}(s))ds\right), \qquad (5.35)$$

where ν is the Gaussian measure ($\int d\nu = 1$) with mean zero concentrated on the paths $\mathbf{r}(.)$ defined on the interval $[0, 1]$ such that $\mathbf{r}(0) = \mathbf{r}(1) = 0$ and characterized by the covariance

$$\int d\nu(\mathbf{r})r^j(s)r^k(s') = \delta^{jk}(s(1 - s')\theta(s' - s) + s'(1 - s)\theta(s - s')). \qquad (5.36)$$

The measure $dW_{(\mathbf{x},\mathbf{y})}^t(w)$ in (5.35) can be defined by its integral on "cylinder functions"

$$\int dW_{(\mathbf{x},\mathbf{y})}^t(w) F(\mathbf{w}(s_1), \mathbf{w}(s_2), ..., \mathbf{w}(s_n)) = \int d\mathbf{x}_1....d\mathbf{x}_n F(\mathbf{x}_1, ..., \mathbf{x}_n)$$
$$\times p(s_1, \mathbf{x}_1 - \mathbf{x})p(s_2 - s_1, \mathbf{x}_2 - \mathbf{x}_1)....p(t - s_n, \mathbf{y} - \mathbf{x}_n). \qquad (5.37)$$

Using the definition (5.37), Eq. (5.36) and the discrete approximation to $\int V$ we can prove that the two lines of Eq. (5.35) coincide. The formula (5.30) at the imaginary time can be analytically continued for a class of analytic potentials and analytic wave functions to the real time

$$((\exp(-itH)\psi)(\mathbf{x}) = \int dW(\mathbf{w}) \exp\left(-i \int_0^t V(\mathbf{x} + \sigma\mathbf{w}(s))ds\right)\psi(\mathbf{x} + \sigma\mathbf{w}(t)),$$
$$(5.38)$$

where

$$\sigma = \sqrt{i} = \frac{1 + i}{\sqrt{2}}.$$

If we insert $\hbar$ and m then $\sigma \to \sqrt{\frac{\hbar}{m}}\sigma$.

We may also consider the action of $\exp(-itH)$ on analytic functions of the form $\psi(a\sigma\mathbf{x})$ where $a \in R$ then

$$((\exp(-itH)\psi)(a\sigma\mathbf{x})$$
$$= \int dW(\mathbf{w}) \exp\left(-i \int_0^t V(\sigma(a\mathbf{x} + \mathbf{w}(s)))ds\right)\psi\left(\sigma(a\mathbf{x} + \mathbf{w}(t))\right), \qquad (5.39)$$

The analytic continuation (5.38) has been investigated first in [71] (see also [50]). The complex coordinates $a\sigma\mathbf{x}$ are discussed in [141] (Sect. 12.7) and [16, 17]. The complex coordinates have been introduced in quantum mechanics (complex scaling) in the theory of resonances (see the reviews [205, 223]). In Sect. 6.1 we discuss potentials which can be treated using Eqs. (5.38)–(5.39).

The formula for the Schrödinger evolution kernel reads

$$\exp(-itH)(\mathbf{x}, \mathbf{y}) = (2\pi it)^{-\frac{3}{2}}$$
$$\times \exp(i\tfrac{(\mathbf{x}-\mathbf{y})^2}{2t}) \int d\nu(\mathbf{r}) \exp\left(-i \int_0^t V(\mathbf{x} + \tfrac{s}{t}(\mathbf{y} - \mathbf{x}) + \sigma\sqrt{t}\mathbf{r}(s))ds\right),$$

For Eq. (5.38) we have to define $w(s)$ for $s < 0$. If $s < 0$ then we use Eq. (5.31) with the Brownian motion $\tilde{w}$ defined for negative s as $\tilde{w}(s) \equiv w'(-s)$, where w' is another Brownian motion (independent of w). The Feynman integral for $-t \geq 0$ reads

$$(\exp(-itH)\psi)(\mathbf{x})$$
$$= \int dW(\mathbf{w'}) \exp\left(i \int_0^{-t} V(\mathbf{x} + \sigma^*\mathbf{w'}(s))ds\right)\psi(\mathbf{x} + \sigma^*\mathbf{w'}(-t)). \qquad (5.40)$$

The formula (5.40) respects the time reflection symmetry of the Schrödinger equation following from the equation

$$-i\partial_t \psi^* = H\psi^*$$

resulting in the formula

$$\psi_t = \left(\exp(itH)\psi^*\right)^*$$

which could have been applied for a definition of $\exp(-iHt)$ for $t \leq 0$.

The integrand in Eq. (5.38) may grow at large w. Then, it is difficult to show the integrability of the exponential factor in Eq. (5.38). We can improve the integrability regularizing the potential as, e.g., for polynomial potentials $V(x) \to V(x) - \epsilon x^{4n+6}$, where n is a (sufficiently large) natural number. Then, we can prove that the integral (5.38) is finite, establish some results as a function of ϵ and finally prove that the limit $\epsilon \to 0$ exists. This can be considered as an ϵ-regularization of the Feynman-Wiener integral (the idea discussed recently in [84]).

At the end of this section we prove that the Feynman formula (5.38) can be considered as a resummation of the Dyson series . Hence, in perturbation expansion Eq. (5.38) holds true for any potential $V(x)$ which has an analytic continuation to $V(\mathbf{x} + \sigma\mathbf{y})$, where $\mathbf{x}, \mathbf{y} \in R$. Let us restrict ourselves to quantum mechanics in one dimension with the Hamiltonian

$$H = -\frac{1}{2}\frac{d^2}{dx^2} + V \equiv H_0 + V.$$

The unitary evolution $\exp(-iH\tau)$ can be expressed in the interaction picture. We write

$$U_\tau = \exp(-iH\tau) = \exp(-iH_0\tau)U_\tau^I. \qquad (5.41)$$

Then,

$$\partial_\tau U_\tau^I = -iV_\tau U_\tau^I, \qquad (5.42)$$

where

$$V_\tau = \exp(iH_0\tau)V\exp(-iH_0\tau). \qquad (5.43)$$

We solve Eq. (5.42) expanding in the potential V (Dyson series)

$$\begin{aligned}
U_\tau &= \exp(-iH_0\tau)U_\tau^I \\
&= \exp(-iH_0\tau)\Big(1 - i\int_0^\tau ds\, V_s - \int_0^\tau ds_2 \int_0^{s_2} ds_1 V_{s_2} V_{s_1} + \cdots\Big).
\end{aligned} \qquad (5.44)$$

We have till the second order term

$$\begin{aligned}
(U_\tau\psi)(x) &= (\exp(-i\tau H_0)\psi)(x) \\
&\quad -i\int_0^\tau ds dx_1 \exp(-iH_0(\tau-s))(x,x_1)V(x_1)\exp(-iH_0s)(x_1,x_2)\psi(x_2)dx_2 \\
&\quad -\int_0^\tau ds_2 \int dx_1 \int dx_2(\exp(-i(\tau-s_2)H_0))(x,x_1)V(x_1) \\
&\quad (\exp(-i(s_2-s_1)H_0))(x_1,x_2)V(x_2)\int_0^{s_2} ds_1(\exp(-is_1 H_0))(x_2,x_3)\psi(x_3)dx_3 + .
\end{aligned} \qquad (5.45)$$

where

$$\begin{aligned}
(\exp(-isH_0))(x,x') &= (2i\pi s)^{-\frac{1}{2}}\exp\left(\frac{i}{2s}|x-x'|^2\right) \\
&\equiv p_F(s, x-x') = p(is, x-x'),
\end{aligned} \qquad (5.46)$$

where $p(s, x-x')$ is defined in Eq. (5.33) (for a three-dimensional space). Going from Eq. (5.44) to Eq. (5.45) we have changed the time integration variables in the first order term $s \to \tau - s$ and in the second order $s_1 \to \tau - s_2$ and $s_2 \to \tau - s_1$. The Feynman path integral (5.38) in the heuristic Feynman notation (5.17) reads

$$\begin{aligned}
(U_\tau\psi)(x) &= \int_{q(0)=x} \mathcal{D}q(.) \\
&\quad \times \exp\left(\frac{i}{2}\int_0^\tau ds(\frac{dq}{ds})^2\right)\exp(-i\int_0^\tau ds V(q(s)))\psi(q(\tau)).
\end{aligned} \qquad (5.47)$$

The lowest order terms in the expansion of the Feynman formula for $U_\tau\psi$ in Eq. (5.47) are

$$\begin{aligned}
< (U_\tau\psi)(x) > &= \int_{q(0)=x} \mathcal{D}q(.)\exp\left(\frac{i}{2}\int_0^\tau (\frac{dq}{ds})^2\right) \\
&\quad \times \Big(1 - i\int_0^\tau ds V(q(s)) - \int_0^\tau ds_2 \int_0^{s_2} ds_1 V_{s_2} V_{s_1} + \cdots\Big)\psi(q(\tau)).
\end{aligned} \qquad (5.48)$$

(5.48) agrees with the Dyson expansion (5.44) (when applied to ψ). In order to show the equality of the expansion in V of the Feynman integral (5.48) and the Dyson expansion (5.44) we calculate the Feynman integral on a dense set of functions. This can be the functions $G(q(s_1), q(s_2), ..., q(s_{n-1}), q(\tau))$ of a finite number of points $0 \leq s_1 \leq s_2 \leq ..... \leq s_{n-1} \leq \tau$ [109, 224]. Then, the Feynman integral of functions of paths starting from $q(0) = x$ and depending on $(q(s_1), q(s_2), ..., q(s_{n-1}), q(\tau))$ is

$$
\begin{aligned}
&\int_{q(0)=x} Dq(.) \exp\left(\tfrac{i}{2} \int_0^\tau ds (\tfrac{dq}{ds})^2\right) G(q(s_1), q(s_2), ..., q(s_{n-1}), q(\tau)) \\
&= \int p_F(s_1, x_1 - x) p_F(s_2 - s_1; x_2 - x_1)....p_F(\tau - s_{n-1}; x_n - x_{n-1}) \\
&\times G(x_1, ..., x_n) dx_1...dx_n
\end{aligned}
\tag{5.49}
$$

for any function $G(x_1, ..., x_n)$ of n-variables. Equation (5.49) is an analytic continuation in time of the corresponding formula (5.32) for the Wiener measure. Equation (5.44) is the same as Eq. (5.48) (with the expectation values calculated according to Eq. (5.49) if in the $(n-1)$th order of Dyson perturbation expansion we change the time integration variables $s_j \to \tau - s_{n-j}$ (as it has been done in Eq. (5.45) when $n = 3$).

We can also prove an analog of the Dyson expansion for the kernel of the Schrödinger evolution

$$
K_\tau(x, x') = \int d\mu_{(x,x')}^\tau(q) \exp(-i \int_0^\tau ds V(q(s))),
\tag{5.50}
$$

where the $d\mu_{(x,x')}^\tau$ integral is defined by its value on the cylinder functions $G(q(s_1), q(s_2), ..., q(s_{n-1}), q(s_n))$. If $0 \leq s_1 \leq s_2 \leq ..... \leq s_n \leq \tau$ then

$$
\begin{aligned}
&\int d\mu_{(x,x')}^\tau(q) G(q(s_1), q(s_2), ..., q(s_n)) \\
&= \int p_F(s_1; x_1 - x) p_F(s_2 - s_1; x_2 - x_1), ...p_F(\tau - s_n; x' - x_n) \\
&\times G(x_1, ..., x_n) dx_1...dx_n.
\end{aligned}
\tag{5.51}
$$

Hence, $d\mu_{(x,x')}^\tau$ may be considered as an analytic continuation $t \to i\tau$ of $dW_{(x,x')}^t$ of Eq. (5.37).

The kernel $K_\tau(x, x')$ (5.50) can be expressed by an analytic continuation of r (5.36) (its Feynman version is derived in [138, 141]) defined as the Gaussian process on the interval $[0, 1]$ with the covariance

$$
< r(s)r(s') > = is(1 - s')\theta(s' - s) + is'(1 - s)\theta(s - s'),
\tag{5.52}
$$

where θ is the Heaviside step function. Let us denote by $d\nu(r)$ the complex Gaussian measure with the covariance (5.52). Then,

$$
\begin{aligned}
K_\tau(x, x'; \mathbf{p}) &= (2i\pi\tau)^{-\frac{1}{2}} \exp\left(\tfrac{i}{2\tau}|x - x'|^2\right) \\
&\times \int d\nu(r) \exp\left(-i\tau \int_0^1 ds V(sx' + (1 - s)x + \sqrt{\tau}r(s))\right)
\end{aligned}
\tag{5.53}
$$

For a mathematically precise meaning of Eq. (5.53) we must return to Eqs. (5.38)–(5.39). Equations (5.52)–(5.53) are just an interpretation of Eqs. (5.36) and (5.39) which have a rigorous functional integral sense (and can be calculated by means of a numerical simulation of the Brownian motion).

5.4   The Stochastic Integral: The Feynman Integral for a Particle in an Electromagnetic Field

The stochastic Ito integral is defined [108, 158, 172, 224] as a limit in $L^2(dW)$ (when $\Delta t_j = t_{j+1} - t_j = \frac{t}{n} \to 0$)

$$F(t) = \int_0^t f(s, w(s)) dw(s) = \lim_{n \to \infty} \sum_{j=1}^n f(t_j, w(t_j))(w(t_{j+1}) - w(t_j)) \quad (5.54)$$

From the definition (because of the independence of increments $\Delta w = w(t_{j+1}) - w(t_j)$) it follows that

$$E\left[\int f dw\right] = 0 \quad (5.55)$$

and

$$E\left[\int_0^t f dw \int_0^{t'} g dw\right] = E\left[\int_0^{min(t,t')} f g ds\right]. \quad (5.56)$$

Equation (5.56) can be proved using the formula

$$E[(w(t_{j+1}) - w(t_j))(w(t_{k+1}) - w(t_k))] = \delta_{jk}(t_{j+1} - t_j). \quad (5.57)$$

The Ito differential of F in Eq. (5.54) is defined as

$$dF = f dw.$$

We can show from the definition of the stochastic integral (first for polynomials and subsequently in general) that

$$dF^2 = 2F f dw + f^2 dt \quad (5.58)$$

and more general

$$d(FG) = dFG + FdG + dFdG \quad (5.59)$$

with the rule $dwdw = dt$ (resulting from Eq. (5.57)). Equation (5.58) also follows from the following sequence of equalities (in the sense of expectation values in $L^2(dW)$)

$$dF^2 = 2FdF + dFdF = 2FF'dw + (F')^2dwdw = 2FF'dw + (F')^2dt,$$

where F' denotes a derivative of F. Similarly, using Taylor expansion we obtain

$$f(w + dw) = f(w) + f'(w)dw + \frac{1}{2}f''dwdw = f(w) + f'(w)dw + \frac{1}{2}f''dt.$$
(5.60)

As follows from Eqs. (5.59)–(5.60) the Ito differential does not satisfy the Leibnitz rules of differentiation. One can define a stochastic integral in a different way (Stratonovitch integral) [158]

$$\begin{aligned}
F(t) &= \int_0^t f(t, w(t)) \circ dw(t) \\
&= \lim_{n \to \infty} \sum_j f\left(\tfrac{1}{2}(w(t_{j+1}) + w(t_j))\right)(w(t_{j+1}) - w(t_j)),
\end{aligned}$$
(5.61)

where $\Delta t_j = t_{j+1} - t_j = \frac{t}{n}$. From the definition (5.61) (the Stratonovitch differential dF of the Stratonovitch integral (5.61) is defined as $dF = f \circ dw$) using the expansion (5.60) one can show that the following relation

$$f \circ dw = fdw + \frac{1}{2}dfdw$$
(5.62)

is satisfied in the sense that we have the equality of integrals

$$\int f \circ dw = \int fdw + \frac{1}{2}\int f'ds,$$

where the first integral on the rhs is the Ito integral and the second one is an ordinary integral of random functions (we used $dfdw = f'dwdw = f'ds$).

As a consequence of Eq. (5.62) the Stratonovitch integral (in contradistinction to (5.59)) satisfies the Leibnitz rule

$$d(FG) = F \circ dG + G \circ dF.$$

Using the stochastic calculus we can prove the Feynman-Kac formula for the Hamiltonian with the electromagnetic field

$$H = \frac{1}{2m}(-i\hbar\nabla + e\mathbf{A})^2 + V.$$
(5.63)

Then,

$$(\exp(-\tfrac{t}{\hbar}H)\psi)(\mathbf{x}) = E\Big[\exp\Big(\tfrac{ie}{\hbar} \int_0^t \mathbf{A}(\mathbf{x} + \sqrt{\tfrac{\hbar}{m}}\mathbf{w}(s)) \circ d\mathbf{w}(s)$$
$$-\tfrac{1}{\hbar} \int_0^t V(\mathbf{x} + \sqrt{\tfrac{\hbar}{m}}\mathbf{w}(s))ds \Big)\psi(\mathbf{x} + \sqrt{\tfrac{\hbar}{m}}\mathbf{w}(t)) \Big]. \tag{5.64}$$

As a consequence of the Leibnitz rule for Stratonovitch integral we obtain the standard Stratonovitch differential equation for $\exp(i \int A \circ dw)$ (see Eq. (5.66) below). Using this differential equation we can show that the phase factor transforms in a covariant way under the gauge transformation (see Chap. 11) so that Eq. (5.64) is covariant with respect to the gauge transformation (see more on the evolution kernel in gauge potentials in Sect. 11.3).

In order to prove Eq. (5.64) it is sufficient to calculate the time derivative of both sides at $t = 0$ because the semigroup property is a consequence of the Markov property of the Brownian motion:

if $F(w(.))$ depends on $w(t)$ with $t > s$ and $G(w(.))$ on $w(\tau)$ with $\tau \le s$ then $F(w(.))$ and $G(w(.))$ are independent random variables
$E[F(w(.))G(w(.))] = E[F(w(.))]E[G(w(.))]$.
Now

$$d(\exp(-\tfrac{t}{\hbar}H)\psi)(\mathbf{x}) = \int dW(\mathbf{w}) \exp\Big(\tfrac{ie}{\hbar} \int_0^t \mathbf{A}(\mathbf{x} + \sqrt{\tfrac{\hbar}{m}}\mathbf{w}(s)) \circ d\mathbf{w}(s) \Big)$$
$$\times \exp\Big(-\tfrac{1}{\hbar} \int_0^t V(\mathbf{x} + \sqrt{\tfrac{\hbar}{m}}\mathbf{w}(s))ds \Big)(-\tfrac{1}{\hbar})V(\mathbf{x} + \sqrt{\tfrac{\hbar}{m}}\mathbf{w}(t))\psi(\mathbf{x} + \sqrt{\tfrac{\hbar}{m}}\mathbf{w}(t))dt$$
$$+ \int dW(\mathbf{w}) \exp(-\tfrac{1}{\hbar} \int_0^t V(\mathbf{x} + \sqrt{\tfrac{\hbar}{m}}\mathbf{w}(s))ds)$$
$$\times \exp\Big(\tfrac{ie}{\hbar} \int_0^t \mathbf{A}(\mathbf{x} + \sqrt{\tfrac{\hbar}{m}}\mathbf{w}(s)) \circ d\mathbf{w}(s) \Big)d\psi(\mathbf{x} + \sqrt{\tfrac{\hbar}{m}}\mathbf{w}(t))$$
$$+ \int dW(\mathbf{w}) \exp(-\tfrac{1}{\hbar} \int_0^t V(\mathbf{x} + \sqrt{\tfrac{\hbar}{m}}\mathbf{w}(s))ds)$$
$$\times d\Big(\exp\Big(\tfrac{ie}{\hbar} \int_0^t \mathbf{A}(\mathbf{x} + \sqrt{\tfrac{\hbar}{m}}\mathbf{w}(s)) \circ d\mathbf{w}(s) \Big)\psi(\mathbf{x} + \sqrt{\tfrac{\hbar}{m}}\mathbf{w}(t)). \tag{5.65}$$

In Eq. (5.65) we used the fact that (see more on it in Sect. 11.3) the Stratonovitch phase factor satisfies the same differential equation as the deterministic one

$$d\Big(\exp\Big(\tfrac{ie}{\hbar} \int_0^t \mathbf{A}(\mathbf{x} + \sqrt{\tfrac{\hbar}{m}}\mathbf{w}(s)) \circ d\mathbf{w}(s) \Big)$$
$$= \tfrac{ie}{\hbar} \exp\Big(\tfrac{ie}{\hbar} \int_0^t \mathbf{A}(\mathbf{x} + \sqrt{\tfrac{\hbar}{m}}\mathbf{w}(s)) \circ d\mathbf{w}(s) \Big)\mathbf{A}(\mathbf{x} + \sqrt{\tfrac{\hbar}{m}}\mathbf{w}(t)) \circ d\mathbf{w}(t). \tag{5.66}$$

We apply the Ito formula (5.60) in order to express $d\psi$ as

$$d\psi(\mathbf{x} + \sqrt{\tfrac{\hbar}{m}}\mathbf{w}(t)) = \sqrt{\tfrac{\hbar}{m}}\nabla\psi d\mathbf{w} + \tfrac{\hbar}{2m}\Delta\psi dt. \tag{5.67}$$

If we let $t = 0$ then using Eq. (5.59), the expression of the Stratonovitch differential (5.66) by the Ito differential from Eq. (5.62), $E[\nabla\psi dw] = 0$ and $dwdw = dt$ we can see that we obtain $-\tfrac{1}{\hbar}H\psi(x)dt$ on the rhs of Eq. (5.65).

5.5   Stochastic Differential Equations

In Eq. (5.66) we have derived a differential equation satisfied by an exponential of a Stratonovitch integral. We may pose a more general problem: consider the Ito differential of a stochastic process $\xi(s)$ (for simplicity we restrict ourselves to $\xi \in R$) as an equation for ξ

$$d\xi(s) = a(s, \xi)ds + b(s, \xi)dw_s, \qquad (5.68)$$

where a, b are continuous functions. This equation can be understood as an integral equation (we follow [108])

$$\xi_{x,t}(s) = x + \int\limits_t^s a(\tau, \xi_{x,t}(\tau))d\tau + \int\limits_t^s b(\tau, \xi_{x,t}(\tau))dw(\tau), \qquad (5.69)$$

where $\xi_{x,t}(t) = x$. Let us define the transition probability

$$P(\xi(t) \in A | \xi(s)) = P(s, \xi(s), t, A) \qquad (5.70)$$

as the probability for ξ to reach the set A in time t if at an earlier moment s it was in $\xi(s)$ ($P(..|\xi)$ denotes the conditional probability under the condition ξ). As the consequence of the facts that $\frac{dw}{d\tau}$ are independent at different τ and the solution of the differential equation (5.68) is uniquely determined by the initial condition we obtain the composition law expressing the Markov property (the future development does not depend on the whole past but only on the initial moment)

$$\int P(s, x, t, dy)P(t, y, \tau, A) = P(s, x, \tau, A), \qquad (5.71)$$

where $0 \le s < t < \tau$. We shall also write $P(s, x, t, dy) \equiv p(s, x, t, y)dy$.

The application of stochastic equations to partial differential equations is based on the formula expressing the solution of the diffusion equation by the solution of the stochastic equation. Let us consider

$$\psi(s, x) = E[\psi(\xi_{x,s}(t))], \qquad (5.72)$$

where ψ is an arbitrary twice continuously differentiable function and $\xi_{x,s}(t)$ is the solution of Eq. (5.69) with the (final) condition $\xi_{x,t}(t) = x$. In terms of the transition function (5.70) we may write Eq. (5.72) in the form

$$\psi(s, x) = E[\psi(\xi_{x,s}(t))] = \int P(s, x, t, dy)\psi(y). \qquad (5.73)$$

We derive the differential equation satisfied by $\psi(s, x)$. We have

$$\psi(s - \epsilon, x) = E[\psi(\xi_{x,s-\epsilon}(t))] = E[\psi(\xi_{x,s-\epsilon}(t))|\xi_{x,s-\epsilon}(s)] = E[\psi(\xi_{x,s-\epsilon}(s))],$$
$$(5.74)$$

where $E[\psi(\xi_{x,s-\epsilon}(t))|\xi_{x,s-\epsilon}(s)]$ denotes the conditional expectation value of the function ψ under the assumption that $\xi_{x,s-\epsilon}(s)$ is known. We used the fact that ξ is the solution of the differential equation (5.68) implying that the Markov property is satisfied. Now

$$\psi(s - \epsilon, x) - \psi(x) = E[\psi(\xi_{x,s-\epsilon}(s)) - \psi(x)]. \qquad (5.75)$$

From the Ito formula (5.60) (Taylor formula now applied to ξ)

$$d\psi(\xi) = \psi' d\xi + \frac{1}{2}\psi'' d\xi d\xi. \qquad (5.76)$$

Using Eq. (5.68) we can write Eq. (5.76) in an integral form

$$\begin{aligned}
\psi(\xi_{s-\epsilon}(s)) - \psi(x) &= \int_{s-\epsilon}^{s} a(\tau, \xi_{x,s-\epsilon}(\tau))\psi'(\xi_{x,s-\epsilon}(\tau))d\tau \\
&+ \frac{1}{2}\int_{s-\epsilon}^{s} b^2(\tau, \xi_{x,s-\epsilon}(\tau))\psi''(\xi_{x,s-\epsilon}(\tau))d\tau \\
&+ \int_{s-\epsilon}^{s} \psi'(\xi_{x,s-\epsilon}(\tau))b(\tau, \xi_{x,s-\epsilon}(\tau)dw(\tau).
\end{aligned} \qquad (5.77)$$

We take the expectation value of Eq. (5.77). Then, we use Eq. (5.55) saying that the expectation value of a stochastic integral with the Brownian motion (the third line in Eq. (5.77)) is zero. There remains in Eq. (5.77) the integral over τ on the interval $[s - \epsilon, s]$. By dividing Eq. (5.77) by ϵ and taking the limit $\epsilon \to 0$ we obtain $-\partial_s\psi(s, x)$ on the lhs. Hence, using $\xi_{x,s}(s) = x$ on the rhs we obtain

$$\partial_s\psi(s, x) = -\frac{1}{2}b^2(s, x)\psi''(x) - a(s, x)\psi'(x). \qquad (5.78)$$

Let us introduce an operator T with $T(t, t) = 1$ as

$$(T(t, s)\psi)(x) = \int P(s, x, t, dy)\psi(y). \qquad (5.79)$$

From the Markov property (5.71) expressed by the transition function we can derive the composition law

$$T(t, s)T(s, \tau) = T(t, \tau) \qquad (5.80)$$

for $t > s > \tau \geq 0$. If the coefficients a and b do not depend on time then $P(s, x, t, dy) = P(t - s, x, dy)$ and $T(t, s) = \exp(\mathcal{M}(t - s))$ where

$$\mathcal{M} = \frac{1}{2}b^2(x)f''(x) + a(x)f'(x). \qquad (5.81)$$

The Feynman-Kac formula takes the form

$$\psi(s, x) = E\left[\exp\left(-\int_s^t V(\tau, \xi_{x,s}(\tau))d\tau\right)\psi(\xi_{x,s}(t))\right] \qquad (5.82)$$

with the final boundary condition $\psi(t, x) = \psi(x)$. As can be checked by a repetition of the argument (5.77) ψ satisfies the equation

$$\partial_s\psi(s, x) = -\frac{1}{2}b^2(s, x)\psi''(s, x) - a(s, x)\psi'(s, x) + V(s, x)\psi(s, x). \qquad (5.83)$$

When we know, that owing to the Markov property, the operator $T(t, s)$ satisfies the composition law (5.80) then in order to derive the differential equation (5.78) it is sufficient to calculate

$$\lim_{\epsilon \to 0} \epsilon^{-1}(T_{t,t-\epsilon} - 1). \qquad (5.84)$$

instead of exploiting Eqs. (5.74) in (5.77).

In Eq. (5.72) we have differentiated over the initial time of the solution of the stochastic equation with a final boundary condition. From the composition law (5.71) it follows that we can inverse the time $s \to t - s$. Then, the stochastic equation (5.68) is replaced by

$$d\xi(s) = a(t - s, \xi)ds + b(t - s, \xi)dw_s. \qquad (5.85)$$

Now, we consider the expectation value

$$\psi(t, x) = E[\psi(\xi_{x,0}(t)], \qquad (5.86)$$

where $\xi_{x,0}(t)$ solves Eq. (5.85) with the initial condition $\xi_{x,0}(0) = x$. It can be shown (similarly as for (5.78)) that (5.86) is the solution of the diffusion equation

$$\partial_t\psi(t, x) = \frac{1}{2}b^2(t, x)\psi''(t, x) + a(t, x)\psi'(t, x) \qquad (5.87)$$

with the initial condition $\psi(0, x) = \psi(x)$.

The Feynman-Kac formula is expressed as

$$\psi(t, x) = E\left[\exp\left(-\int_0^t V(\xi_{x,0}(t - \tau))d\tau\right)\psi(\xi_{x,0}(t))\right]. \qquad (5.88)$$

ψ satisfies the equation

$$\partial_t\psi(t, x) = \frac{1}{2}b^2(t, x)\psi''(t, x) + a(t, x)\psi'(t, x) - V(t, x)\psi(t, x). \qquad (5.89)$$

In [96] the time-inhomogeneous equation (5.68) is treated as a system of homogeneous equations by an introduction of an auxiliary variable ϑ

$$d\xi(s) = a(\vartheta, \xi)ds + b(\vartheta, \xi)dw(s),\tag{5.90}$$

$$d\vartheta = -ds.\tag{5.91}$$

For a homogeneous system (5.90)–(5.91) we can define the semigroup
$E[\psi(\vartheta(s), \xi(s))]$ (where $\vartheta(s) = t - s$) which is expressing the results (5.86)–
(5.87). This is a method similar to Howland's treatment of time-dependent Hamiltonians discussed in Sect. 5.2 (see Eq. (5.29)). The dynamical system (5.90)–(5.91)
gives an alternative proof of Eqs. (5.86)–(5.89).

The Ito stochastic differential equations (5.68) can be rewritten in the form of
Stratonovitch equations. Such equations can be treated like ordinary differential
equations (because the standard Leibnitz rules of differentiation apply). In subsequent
chapters in explicit solutions we restrict ourselves to a simple linear equation

$$d\xi = a(t - s)\xi(s)ds + b(t - s)dw(s).\tag{5.92}$$

It has the solution (with $\xi(0, x) = x$)

$$\xi(s, x) = \exp\left(\int_0^s a(t - \tau)d\tau\right)x + \int_0^s \exp\left(\int_\tau^s a(t - u)du\right)b(t - \tau)dw(\tau).$$

$$\tag{5.93}$$

5.6   Exercises

5.1 Using the Riemannian sum approximation for $\int^t V$ in the formula (5.35) and the
cylindrical integral (5.37) show the formula (5.14) for the Trotter approximation
(this proves that the discretization of the Wiener integral gives the Trotter product
formula).
5.2 Show the same as in Exercise 5.1 for the Feynman-Wiener formula (5.38).
5.3 Prove Eq. (5.53) using the Riemannian finite sum approximation as indicated at
the end of Sect. 5.3 using the methods suggested in Exercises 5.1 and 5.2.
5.4 Prove (5.58) in the form

$$F^2(w_t) - F^2(w_0) = 2\int_0^t F\,dF + \int_0^t (F')^2 ds$$

directly from the definition (5.54). Check the above mentioned formula calculating multi-time correlation functions for both side of the formula for polynomial
F.

5.5 Check the Feynman-Kac formula (5.35) and (5.38) for $V(t, x) = a(t)x$ by means of an explicit solution of the Schrödinger equation and an explicit calculation of the Wiener-Feynman integral (5.38)

Hint: for calculations use the Fourier representation of the initial wave function as $\psi(x) = \int \psi(p) \exp(ipx)dp$ (the Schrödinger equation can easily be solved in the momentum representation).

5.6 Show at an arbitrary order that the Dyson series (5.45) and an expansion of the Feynman integral (5.48) coincide.

5.7 Prove Eq. (5.66)

5.8 Apply in the formulas (5.86)–(5.87) the expression (5.93) for $b = 1$ and $a = -\omega x$ (the Ornstein-Uhlenbeck process). Relate this process to the harmonic oscillator at imaginary time. Calculate the explicit solution for various initial conditions ψ.

Chapter 6
Feynman Integral in Terms of the Wiener Integral

Abstract The Feynman integral (in the way formulated by Feynman) has no rigorous mathematical meaning as shown by Gelfand and Yaglom. An imaginary time version can be considered as a way to treat the path integral. However, for some problems in quantum mechanics, e.g., the scattering and interference, we must work with a real time. Moreover, for potentials which are unbounded from below the imaginary time version is not applicable. For these reasons a real time formulation of the mathematical theory of the Feynman integral is necessary. In this chapter we apply the method of an analytic continuation of the Wiener integral to polynomial potentials. Then, we consider the Feynman integral for the class of functions which are Fourier-Laplace transforms of a complex measure. The method of an analytic continuation of the Wiener integral is applied to QFT when the potential perturbs the paths of a free field (an infinite dimensional oscillator). We show that for trigonometric and exponential potentials in two-dimensional QFT the rigorous methods (usually exploited in Euclidean field theory) based on the Feynman-Kac formula can be applied directly in the real time.

The Feynman integral (in the way formulated by Feynman) has no rigorous mathematical meaning as shown by Gelfand and Yaglom [105]. An imaginary time version can be considered as a way to treat the path integral. However, for some problems in quantum mechanics, e.g., the scattering and interference, we must work with real time. Moreover, for potentials which are unbounded from below (the simplest example is the potential $V = -v^2x^2$) the imaginary time version is not applicable. For these reasons a real time formulation of the mathematical theory of the Feynman integral is necessary. In this chapter we apply the method of an analytic continuation of the Wiener integral initiated in Sect. 5.3 in application to polynomial potentials (Sect. 6.1). In Sect. 6.2 we extend this method to define the Feynman integral for the class of functions which are Fourier-Laplace transforms of a complex measure. Then, the method is applied to QFT when the potential perturbs the paths of a free field (an infinite dimensional oscillator). We show that for trigonometric and exponential potentials in two-dimensional QFT the rigorous methods (usually exploited in Euclidean field theory) based on the Feynman-Kac formula can be applied directly in the real time (polynomial interactions are discussed in Sect. 8.7).

Returning to the past. The aim to formulate the Feynman integral as a rigorous mathematical tool absorbed many mathematicians. It was known that in the original Feynman formulation there was no measure on the paths [105]. Then, there appeared various formulations which have some virtues and some drawbacks. There is the Fresnel integral approach mentioned in [160] and developed in [8]. It applies to potentials and wave functions which are Fourier transforms of a measure. However, some intuitive aspects of the Feynman integral are lost in this approach. Another approach initiated in [50, 71] relies on the well-defined Wiener measure but it is using complex paths in expectation values. The method can be applied solely to some analytic potentials. It has the virtue that it preserves most formal properties of the Feynman integral and (as based on stochastic processes) is applicable for numerical simulations.

6.1   Feynman-Wiener Integral for Polynomial Potentials

As discussed in Sect. 5.3 an analytic continuation of the Wiener integral leads to a representation of the unitary evolution which has the form of a Feynman integral. Its perturbation expansion coincides with the Dyson expansion as shown in Sect. 5.3. So if the Feynman integral exists it is a legitimate resumation of the perturbation series. The difficulty is that in Eq. (5.38) the term

$$\exp\left(-i\int\limits_0^t V(x+\sigma w(s))ds\right) \tag{6.1}$$

in general is an unbounded function of w. Hence, its expectation value may be infinite. The function (6.1) is explicitly bounded in w for polynomial potentials with the highest order term gx^{2n} if

$$-ig\sigma^{2n} = -gi^{n+1} < 0. \tag{6.2}$$

So, if $n + 1 = 2k$, where k is a natural number, then the condition (6.2) is $-g(-1)^k < 0$. Hence, k should be even if $g > 0$ and k odd if $g < 0$. The simplest case is the potential $V = -v^2 x^2$ corresponding to $k = 1$. These potentials with bounded exponentials (6.1) are rather special. However, we may consider arbitrary polynomial potentials treating the addition of the potential gx^{2n} (with a sufficiently large n) as a regularization (a similar method of regularization is suggested in [84]). After establishing some properties of the solutions of the Schrödinger equation we can study the limit $g \to 0$.

As an example we consider $g > 0$

$$V(x) = gx^6 + \lambda x^4 + \frac{1}{2}\omega^2 x^2.$$

Then

$$\left| \exp\left(-i \int_0^t ds V(x + \sigma w(s)) \right) \right|$$
$$= \exp\left(-\int_0^t ds (gw(s)^6 + p(x, w(s))) \right) \leq K(t, x),$$

where $p(x, w(s))$ is a real polynomial of a fifth order and $K(t, x)$ is a function which does not depend on w. The bound holds true because $gw^6 + p(x, w) \geq -\kappa(x)$ for a certain function $\kappa(x)$. In such a case the rhs of Eq. (5.38) is integrable if

$$E[|\psi(x + \sigma w(t))|] < \infty.$$

This is the case for functions ψ which are Fourier-Laplace transforms of a measure discussed in the next section.

The analytic continuation of coordinates in Eq. (5.39) leads to a larger class of potentials which can be treated by means of the Feynman-Wiener integral. If in Eq. (6.1) $V(x + \sigma w(s))$ is replaced by $V(\sigma(ax + w(s))$ then

$$E\left[\left| \exp\left(-i \int_0^t V(\sigma ax + \sigma w(s)) ds \right) \right| \right] < \infty$$

for polynomials of the form

$$V(x) = \alpha x + \sum_{n=1}^{N} c_{2n} x^{2n}, \tag{6.3}$$

where α is an arbitrary real number, n is a natural number. If $n = 2k$ then σ^{4k} is real and $|\exp(i(a\sigma w)^{2n})| = 1$, so c_{4k} can be an arbitrary real number. If n is odd and $ic_{2n}\sigma^{2n} > 0$ then $\exp(-ic_{2n}\sigma^{2n}) < 1$. Hence, in the Feynman-Kac formula (5.39) the exponentil is bounded. These conditions are satisfied for example for the potential $V(x) = c_2 x^2 + c_4 x^4$ where $c_2 < 0$ and c_4 can be either positive or negative. In general, if $N = 2k + 1$ then we need for integrability of the expression in the Feynman-Kac formula $c_{2N}\sigma^{4k+2}i > 0$. Hence, $c_{2N}(-1)^k < 0$. If the term with the highest power $c_{2N}x^{2N}$ of the potential (6.3) is negative then the Hamiltonian is not essentially self-adjoint [203]. In such a case the time evolution is not determined in the unique way by the Hamiltonian. We must choose the self-adjoint extension. The Feynman path integral may be considered as a method for the choice of the extension as suggested by Nelson [190]. The scalar product will be preserved by evolution $(\psi_t, \psi_t) = (\psi_0, \psi_0)$ if $\int dx \partial_x (\psi_t^* \partial_x \psi_t - \psi_t \partial_x \psi_t^*) = 0$. The use of complex coordinates and complex scaling have been developed in a description of resonances in quantum mechanics (see the reviews in [205, 223]). The scaling $x \to a\sigma x$ is considered in [141] (Sect. 12.7) and in [16, 17, 186]. The path integral (5.38)–(5.39) in the limit $\hbar \to 0$ leads to a semiclassical expansion with complex trajectories [139]. The complex scaling (5.39) can be used in an approach to QFT with negative potentials which are of interest in cosmology [195] as we discuss in Sect. 8.7.

6.2 Feynman-Wiener Integral for Potentials Which are Fourier-Laplace Transforms of a Measure

We consider the Schrödinger equation (with $m = 1$)

$$\partial_t \psi = \hat{H}\psi, \tag{6.4}$$

where

$$\hat{H} = i\frac{\hbar}{2}\frac{d^2}{dx^2} - \frac{i}{\hbar}V(x) = \hat{H}_0 - \frac{i}{\hbar}V(x) \tag{6.5}$$

and (where $a \in R$)

$$V(x) = g\int d\mu(a)\exp(i\alpha a x), \tag{6.6}$$

$$\psi(x) = \int d\nu(a_0)\exp(i\alpha a_0 x). \tag{6.7}$$

We admit either $\alpha = 1$ or $\alpha = i$. Note that $\mu(a) = \mu(-a)^*$ for $\alpha = 1$ and $\mu(a) = \mu(a)^*$ for $\alpha = i$ if the potential is to be real. If $\mu(a) = \mu_1(a) + i\mu_2(a)$, where μ_j are real, then we define the norm $|\mu| = |\mu_1| + |\mu_2|$, where $|\mu_j|$ is the variation of the measure on R so that $|\int f d\mu| \leq |\mu|\sup|f|$. The potentials which are Fourier transforms of a measure (with $\alpha = 1$) have been studied in the approach to the Feynman integral in terms of the Fresnel integral in ref. [10].

We can solve the Schrödinger equation (6.4) perturbatively by means of the Dyson (Duhamel) expansion (as in Eq. (5.44))

$$\psi_t = \lim_{N\to\infty}\sum_{k=0}^{N}\psi_t^k \equiv \exp(\hat{H}_0 t)\psi - \frac{i}{\hbar}\int_0^t ds\,\exp((t-s)\hat{H}_0)V\exp(s\hat{H}_0)\psi$$
$$-\hbar^{-2}\int_0^t ds\int_0^s ds'\,\exp((t-s)\hat{H}_0)V\exp((s-s')\hat{H}_0)V\exp(s'\hat{H}_0)\psi + \cdots \tag{6.8}$$

By direct differentiation we can check that if the series (6.8) is uniformly convergent, so that we can exchange differentiation with the (infinite) sum, then the sum of the series is the solution of the Schrödinger equation.

We wish to express the Feynman integral by the Wiener measure defined as the Gaussian measure with mean zero and the covariance

$$E[w_t w_s] = min(t, s), \tag{6.9}$$

where $w_0 = 0$ and $t, s \geq 0$.

We define the stochastic process

$$q_s(x) = x + \sigma w_s, \tag{6.10}$$

where

$$\sigma = \sqrt{i\hbar} \equiv \frac{1}{\sqrt{2}}(1+i)\sqrt{\hbar}. \tag{6.11}$$

It satisfies the initial condition $q_0 = x$. We mentioned in Sect. 5.3 that

$$(\exp(t\hat{H}_0)\psi)(x) = E[\psi(q_t(x))]. \tag{6.12}$$

We consider here functions ψ of the form (6.7). In such a case we can check the formula (6.12) calculating explicitly the Gaussian integral (see Eq. (1.11))

$$E[\exp(i\alpha a_0 q_s)] = \exp\left(E[i\alpha a_0 q_s)] - \frac{1}{2}\alpha^2 a_0^2 E\left[(q_s - E[q_s])^2\right]\right), \tag{6.13}$$

where

$$E[q_s] = x. \tag{6.14}$$

Let us define

$$S_N(-\tfrac{i}{\hbar}\int_0^t V) \equiv \exp_N(-\tfrac{i}{\hbar}\int_0^t V)$$
$$= \sum_{k=0}^N (-\tfrac{i}{\hbar})^k \int_0^t ds_k ...... \int_0^{s_2} ds_1 V(q_{s_k})....V(q_{s_1}). \tag{6.15}$$

We have the point-wise limit

$$lim_{N\to\infty}\exp_N\left(-\frac{i}{\hbar}\int_0^t V\right) = \exp\left(-\frac{i}{\hbar}\int_0^t ds V(q_s(x))\right). \tag{6.16}$$

We can show that the series (6.8) and

$$E\left[\exp_N\left(-\frac{i}{\hbar}\int_0^t V\right)\psi(q_t(x))\right] \tag{6.17}$$

coincide term by term in an expansion in V. This has already been shown by means of direct calculations at the end of Sect. 5.3 but let us still give here another proof using the Markov property. The zero order terms are equal according to Eq. (6.12). Then, let us note that

$$(\exp(t\hat{H}_0)\psi)(x) = (\exp((t-s)\hat{H}_0)\exp(s\hat{H}_0)\psi)(x) \tag{6.18}$$

which according to Eq. (6.12) gives

$$E[\psi(q_t(x))] = E[\psi\Big(q_s(q_{t-s}(x))\Big)]. \tag{6.19}$$

Hence

$$q_t(x) \equiv q_s(q_{t-s}(x)) \tag{6.20}$$

as random variables (the composition law of random paths). We outline the idea of the (Markov) proof that (6.8) and (6.17) coincide at each order of the expansion on the basis of an example of the first order term. In Eq. (6.17) this is

$$-\frac{i}{\hbar} \int_0^t ds\, E[V(q_s(x))\psi(q_t(x))] = -\frac{i}{\hbar} \int_0^t ds\, E[V(q_{t-s}(x))\psi(q_s(q_{t-s}(x)))],$$
$$\tag{6.21}$$

where we have first changed the integration time $s \to t - s$ and subsequently applied the Markov property (6.20). It can be seen that (6.21) coincides with the first order term in Eq. (6.8) equal to

$$-\frac{i}{\hbar} \int_0^t \exp((t - s)\hat{H}_0)V \exp(s\hat{H}_0)\psi. \tag{6.22}$$

To show that (6.21) and (6.22) coincide we use the representation $(\exp(t\hat{H}_0)\phi)(x) = E[\phi(q_t(x))]$ for both $\phi = \psi$ and $\phi = V \exp(s\hat{H}_0)\psi$. This procedure (of the change of time integration variable and subsequent repeated use of the Markov property) can be applied to each term in the expansions (6.8) and (6.17).

As a consequence, the estimates of the series (6.8) and (6.17) are equivalent. We can show by a calculation of the expectation value (6.17) that

(i) If $\alpha = 1$, $\int d|\mu| < \infty$ and $\int d|v| < \infty$ then the series (6.17) is absolutely convergent. The solution satisfies the Schrödinger equation (6.4) and the bound

$$|\psi_t(x)| \leq \exp\left(\frac{t}{\hbar} \int d|\mu|\right) \int d|v|. \tag{6.23}$$

(ii) if $\alpha = i$ then the series is absolutely convergent, the sum satisfies the Schrödinger equation (6.4) and the bound

$$|\psi_t(x)| \leq |\psi(x)| \exp\left(\frac{t}{\hbar}|V(x)|\right). \tag{6.24}$$

In order to prove the convergence of the perturbation series (6.17) (the same as (5.48)) we calculate the expectation value of the N-th order term. When we apply the Fourier-Laplace representation (6.6)–(6.7) then we can see that the Nth order term is of the form (we skip the time integration in front of the Nth order term)

$$\int dv(a_0)\Pi_j d\mu(a_j) \exp\left(-\tfrac{1}{2}\alpha^2\sigma^2 \sum_{\gamma,\beta} a_\gamma a_\beta E\left[w_{s_\gamma} w_{s_\beta}\right]\right)$$
$$\times \exp(\sum_{\gamma,\beta} f(a_\gamma, a_\beta)) \equiv \int d\mu dv \exp(F)\exp(f), \tag{6.25}$$

where the indices γ, β are either 0 or j (a_j coming from the potential and a_0 from ψ). We have in Eq. (6.25) $|\exp F| = 1$ for $\alpha = 1$ as well as for $\alpha = i$ (because $\sigma^2 = i\hbar$). The second multiplicative term in the second line of Eq. (6.25) is

$$\exp f(a_j, a_k) = \exp(ia_j\alpha x + ia_k\alpha x),$$

$$\exp f(a_0, a_j) = \exp(ia_j\alpha x + ia_0\alpha x).$$

It is sufficient to bound the $\exp(f)$ term in the series. If $\alpha = 1$ then $\cdot f$ is purely imaginary and $|\exp f| = 1$ leading to the bound (6.23) in (i) as the Nth order term can be bounded by $(N!)^{-1}t^N(\int d|\mu|)^N \int d|\nu|$ where the factor $(N!)^{-1}$ comes from the time-ordered integral.

For $\alpha = i$ the N-th order term is bounded by

$$\left| \int \Pi_j d\mu(a_j) d\nu(a_0) \exp(f) \right| (N!)^{-1}$$

which is bounded by $|\psi||V|^N (N!)^{-1}$ leading to the bound (6.24).

We consider next a perturbation of the harmonic oscillator

$$V = \frac{1}{2}\omega^2 x^2 + \tilde{V} - \frac{1}{2}\hbar\omega, \tag{6.26}$$

where

$$\tilde{V}(x) = g \int d\mu(a) \exp(i\alpha a x).$$

We are interested in perturbations (6.26) of the oscillator from the point of view of an application to QFT where the interacting quantum field can be considered as a perturbation of the (oscillatory) free field. We cannot use a perturbation expansion in V (6.26) around the free Hamiltonian (even if $\tilde{V}$ is a bounded function) because the perturbation series for the oscillator evolution is convergent only till $|t| < \frac{\pi}{\omega}$ (as is evident from the focal points of the harmonic oscillator present in its evolution kernel, see Eq. (7.23) in Chap. 7 and [184, 187]). For this reason we must treat the oscillator separately.

In the oscillatory case (6.26) it is convenient to write the initial condition as

$$\psi = \psi^g \chi \equiv \left(\frac{\pi\hbar}{\omega}\right)^{-\frac{1}{4}} \exp\left(-\frac{\omega x^2}{2\hbar}\right)\chi, \tag{6.27}$$

where ψ^g is the ground state solution of the Schrödinger equation with $\tilde{V} = 0$.

Then, the solution of the Schrödinger equation has the form

$$\psi_t = \psi^g \chi_t, \tag{6.28}$$

where χ_t satisfies the equation

$$\partial_t \chi = \left(\tilde{H}_0 - \frac{i}{\hbar} \tilde{V} \right) \chi \qquad (6.29)$$

with

$$\tilde{H}_0 = \frac{i\hbar}{2} \frac{d^2}{dx^2} - i\omega x \frac{d}{dx}. \qquad (6.30)$$

We consider the stochastic equation

$$dq_s == -i\omega q ds + \sigma dw_s \qquad (6.31)$$

with the solution (where x is the initial condition at t_0)

$$q_s(x) = \exp(-i\omega(s - t_0))x + \sigma \int_{t_0}^{s} \exp(-i\omega(s - t))dw_t.$$

Using $E[(\int f dw_s)^2] = \int f^2 ds$ (Eq. (5.56)) we calculate

$$\int dx |\psi^g(x)|^2 E[q_t(x)q_{t'}(x)] = \frac{\hbar}{2\omega} \exp(-i\omega|t - t'|).$$

The rhs of this equation is the expectation value in the ground state of the time-ordered product of Heisenberg picture position operators of the harmonic oscillator.

When $t_0 = -\infty$ and we neglect $\exp(i\omega t_0) \simeq 0$ then

$$q_s(x) = \sigma \int_{-\infty}^{s} \exp(-i\omega(s - t))dw_t.$$

is independent of x and (again with $\exp(i\omega t_0) \simeq 0$)

$$E[q_t(x)q_{t'}(x)] \simeq \frac{\hbar}{2\omega} \exp(-i\omega|t - t'|).$$

We encounter already the approximation of oscillating terms as zero at a large time in the derivations in the interaction picture in Chap. 3. It can be justified by the Lebsgues lemma (Eq. (3.27)). We cannot apply it here but it can be used in quantum field theory in the next section when $\exp(i\omega(k)t_0) \simeq 0$ is integrated over k.

The first order equation (6.31) is related to a stochastically perturbed oscillator equation. Differentiating Eq. (6.31) we obtain a second order equation for a complex coordinate q_s

$$\partial_s^2 q_s + \omega^2 q_s = -i\sqrt{i\hbar}\omega \partial_s w_s - i\sqrt{i\hbar}\partial_s^2 w_s.$$

Eq. (6.31) can also be rewritten in terms of real components (q_1, q_2) of $q = q_1 + iq_2$. If we introduce $q_+ = q_1 + q_2$ and $q_- = q_1 - q_2$ then these components satisfy the equations

$$\partial_s q_- = \omega q_+$$

and

$$\partial_s^2 q_- + \omega^2 q_- = \omega\sqrt{2\hbar}\partial_s w \qquad (6.32)$$

with the solution

$$q_-(t) = \cos(\omega(t - t_0))q_- + \sin(\omega(t - t_0))q_+ + \sqrt{2\hbar}\int\limits_{t_0}^{t} \sin(\omega(t - s))dw_s.$$

We are going to prove that the solution of Eq. (6.29) can be expressed as an average over the oscillatory paths (6.31)

$$\chi_t(x) = E\left[\exp\left(-\frac{i}{\hbar}\int\limits_{t_0}^{t} \tilde{V}(q_s)ds\right)\chi(q_t(x))\right], \qquad (6.33)$$

where $\tilde{V}$ has the representation (6.6) and χ the representation (6.7)
$\chi = \int d\nu(a_0)\exp(ia_0 a_0 x)$. We do not prove that $\left|\exp\left(-\frac{i}{\hbar}\int_{t_0}^{t} \tilde{V}(q_s)ds\right)\right|$ is integrable. For this purpose we would need some cutoffs (as in [15, 137]) which subsequently are removed in the perturbation series. We show that the perturbation series of the formula (6.33) in powers of $\tilde{V}$ is absolutely convergent. Similarly as for Eqs. (6.4)–(6.5) we can show that the Dyson expansion (6.8) and the expansion of the exponential (6.33) in powers of $\tilde{V}$ coincide (now there is the oscillator Hamiltonian as H_0 in the Dyson series). In order to prove the convergence of the perturbation series we calculate the expectation value of the Nth order term. When we apply the Fourier-Laplace representation of χ and $\tilde{V}$ then we can see that the N-th order term is of the form (we skip the time integration in front of the Nth order term)

$$\int d\nu(a_0)\Pi_j d\mu(a_j)\exp(\textstyle\sum_{\gamma,\beta} f(a_\gamma, a_\beta))\exp\left(-\tfrac{1}{2}\sigma^2\alpha^2\right.$$
$$\times \textstyle\sum_{\gamma,\beta} a_\gamma a_\beta E\left[\int_{t_0}^{s_\beta} \exp(-i\omega(s_\beta - s))dw_s \int_{t_0}^{s_\gamma} \exp(-i\omega(s_\gamma - s'))dw_{s'}\right])$$
$$= \int d\nu(a_0)\Pi_j d\mu(a_j)\exp(\textstyle\sum_{\gamma,\beta} f(a_\gamma, a_\beta)) \qquad (6.34)$$
$$\times \exp\left(-\tfrac{1}{2}\sigma^2\alpha^2 \textstyle\sum_{\gamma,\beta} a_\gamma a_\beta \int_{t_0}^{min(s_\gamma, s_\beta)} \exp(-i\omega(s_\beta + s_\gamma - 2s))ds\right)$$
$$\equiv \int d\mu d\nu \exp(f)\exp(F),$$

where the indices γ, β are either 0 or j (coming from χ and the potential $\tilde{V}$). We applied Eq. (5.56) in (6.34) when calculating the expectation value of the stochastic integral. We have

$$f(a_j, a_k) = i\alpha a_j x \exp(-i\omega(s_j - t_0)) + i\alpha a_k x \exp(-i\omega(s_k - t_0)) \qquad (6.35)$$

and

$$f(a_0, a_k) = i\alpha a_0 x \exp(-i\omega t)) + i\alpha a_k x \exp(-i\omega(s_k - t_0)). \tag{6.36}$$

The term $|\exp F|^2$ in Eq. (6.34) is

$$\left| \exp\left(-\tfrac{1}{2}\sigma^2\alpha^2 \sum_{\gamma\beta} a_\beta a_\gamma \int_{t_0}^{min(s_\beta,s_\gamma)} \exp(-i\omega(s_\beta + s_\gamma - 2s))ds \right) \right|^2$$
$$= \exp\left(-\tfrac{\hbar\alpha^2}{\omega} \sum_{\beta,\gamma} a_\beta a_\gamma \Big(\cos(\omega(s_\beta - s_\gamma)) - \cos(\omega(s_\beta + s_\gamma - 2t_0)) \Big) \right). \tag{6.37}$$

Let us note that in the Fourier case ($\alpha = 1$) we have

$$-\frac{\hbar\alpha^2}{\omega} \sum_{\beta,\gamma} a_\beta a_\gamma \cos(\omega(s_\beta - s_\gamma)) \leq 0.$$

The second term on the rhs in the second line of Eq. (6.37) (depending on t_0) has an indefinite sign. Hence, it can diverge if the range of integration over a is infinite. There is still the integral of $\exp(f(a_\beta, a_\gamma))$ depending on oscillator terms $\exp(-i\omega t_0)$. As $\Re f(a_\beta, a_\gamma)$ can grow as a function of a the range of a in $d\mu(a)$ must be bounded if the perturbation series for the solution (6.33) is to be convergent. The terms (6.35)–(6.37) are bounded for both $\alpha = 1$ and $\alpha = i$ if the range of a_α is bounded.

Let us consider the Gell-Mann-Low formula for the potentials (6.26). Let H be the Hamiltonian with the potential (6.26). Then, according to the Gell-Mann-Low formula (Eq. (5.5)) the correlation functions of the time-ordered position operators in the Heisenberg picture in the physical ground state Φ_0 can be calculated as expectation values in the ground state $|0>$ of the harmonic oscillator. By formal arguments as applied for quantum fields in Eq. (3.72) these correlation functions can be expressed by the oscillator stochastic process. So, we calculate

$$< \Phi_0 | T\Big(\exp(i\alpha a x_{t_1})... \exp(i\alpha a x_{t_n}) \Big) | \Phi_0 >$$
$$= Z^{-1} \int dx |\psi^g(x)|^2$$
$$E\Big[\exp(i\alpha a q_{t_1}(x))... \exp(i\alpha a q_{t_n}(x)) \exp\Big(-\tfrac{i}{\hbar} \int_{-\infty}^{\infty} \tilde{V}(q_s(x))ds \Big) \Big], \tag{6.38}$$

where

$$Z = \int dx |\psi^g(x)|^2 E\Big[\exp\Big(-\frac{i}{\hbar} \int_{-\infty}^{\infty} \tilde{V}(q_s(x))ds \Big) \Big].$$

x_t is the position operator in the Heisenberg picture, $T(..)$ denotes the time ordered product, $q_t(x)$ is the solution of the stochastic equation (6.31). We obtain a convergent expansion with a finite range of a. In quantum field theory ω is a function of momentum. An integration over the momentum enables an application of the Lebesques lemma to oscillatory integrals so that the Gell-Mann-Low perturbation series with an infinite range of a can be applicable.

We have shown in this section using the Feynman integral in the class of functions which are Fourier-Laplace transforms of a measure with a bounded range of a that the perturbation series is convergent. The statement can be expressed as the existence of the limit for an expectation value of a functional G

$$< G >= \lim_{n \to \infty} \int dW(q)\mathcal{F}_n(q)G(q)$$

with the Wiener measure dW and a certain approximation (or regularization) $\mathcal{F}_n$ of the Feynman integral. We do not know whether the limit can be moved under the integral sign. The rhs of this equation can be considered as a definition of the Feynman integral. In general, we do not know whether the function inside the square brackets in Eq. (6.17) is integrable. We could (as we did in [15, 137]) regularize the potential, show that the functional integral exists for the regularized integral, expand it in a perturbation series and subsequently remove the regularization. These methods can be considered as an alternative to the discrete (finite dimensional) path integral definition due originally to Feynman.

6.3 Functional Integration in Terms of Oscillatory Paths in QFT

In this section we apply the oscillatory paths of the previous section to quantum field theory. In the functional representation of Sect. 2.5 the Hamiltonian for quantum fields (Eqs. (2.60) and (3.2)) is of the form (6.29) as a perturbation of the harmonic oscillator (where the oscillator frequency ω depends on the wave vector $\mathbf{k}$). The aim is to extend the representation of the solution (6.33) of the Schrödinger equation to QFT which involves a generalization of the Brownian motion (the Wiener process) to an infinite number of dimensions .

We begin with the canonical field theory with the Hamiltonian (in a formal Hilbert space $L^2(d\Phi)$ of Sect. 2.5)

$$H = \frac{1}{2} \int d\mathbf{x}\left(\Pi^2 + (\omega\Phi)^2\right) + \int d\mathbf{x}V(\Phi) \equiv H_0 + \int d\mathbf{x}V(\Phi), \qquad (6.39)$$

where

$$\omega = \sqrt{-\Delta + m^2}$$

and

$$[\Phi(\mathbf{x}), \Pi(\mathbf{y})] = i\hbar\delta(\mathbf{x} - \mathbf{y}). \qquad (6.40)$$

Let ψ^g be the ground state (time-independent) solution of the Schrödinger equation

$$i\hbar\partial_t\psi = H_0\psi. \qquad (6.41)$$

We express the solution of the Schrödinger equation (6.41)) with the initial condition

$$\psi = \psi^g \chi \tag{6.42}$$

in the form $\psi_t = \psi^g \chi_t$. Then, χ_t satisfies the equation

$$\partial_t \chi = -\frac{i}{2\hbar} \int d\mathbf{x} \Big(\Pi^2 + 2(\Pi \ln \psi^g)\Pi \Big) \chi \equiv \hat{H}_0 \chi, \tag{6.43}$$

where

$$\Pi(\mathbf{x}) = -i\hbar \frac{\delta}{\delta\Phi(\mathbf{x})}. \tag{6.44}$$

$\hat{H}_0$ is the transformed Hamiltonian (2.60) of Sect. 2.5 in the Hilbert space $L^2(d\mu_0)$ defined by means of the Gaussian measure $d\mu_0 = |\psi^g|^2 d\Phi$ with the rigorous definition of Sect. 1.3. So, the transformation (6.42)) and a subsequent transformation of paths can be considered as a necessary procedure in order to pass to the mathematically well-defined Hilbert space $L^2(|\psi^g|^2 d\Phi)$.

The ground state solution of Eq. (6.41) is

$$\psi^g = Z^{-1} \exp\Big(-\frac{1}{2\hbar} \Phi\omega\Phi \Big), \tag{6.45}$$

where Z is chosen in such a way that in finite dimensional approximations $\int d\Phi |\psi^g|^2 = 1$. Then, in Eq. (6.43)

$$\Pi \ln \psi^g = i\omega\Phi. \tag{6.46}$$

Equation (6.43) is a diffusion equation in infinite dimensional spaces with an imaginary diffusion constant. In order to obtain a solution of such equations we define the Wiener process in an infinite dimensional space [122, 172] as the Gaussian process with the covariance

$$E\Big[W_t(\mathbf{x}) W_s(\mathbf{y}) \Big] = min(t,s)\delta(\mathbf{x} - \mathbf{y}). \tag{6.47}$$

Its Fourier transform $W_t(\mathbf{k})$ has the covariance

$$E\Big[W_t(\mathbf{k}) W_s(\mathbf{k}') \Big] = min(t,s)\delta(\mathbf{k} + \mathbf{k}').$$

In the framework of Sect. 1.3 the Wiener process is defined by the Fourier transform of the Wiener measure $d\mu(W)$ as

$$\int d\mu(W) \exp\Big(i \int ds d\mathbf{x} f(s, \mathbf{x}) W_s(\mathbf{x}) \Big)$$
$$= \exp\Big(-\tfrac{1}{2} \int ds dt d\mathbf{x} f(s, \mathbf{x}) f(t, \mathbf{x}) min(t, s) \Big). \tag{6.48}$$

We regularize the Hamiltonian (6.39) so that when expressed in Fourier space the heuristic free Hamiltonian takes the form (with the vacuum energy subtracted)

$$H_0^\kappa = \frac{1}{2} \int d\mathbf{k} \Big(\Pi(\mathbf{k})\Pi(-\mathbf{k}) + \kappa(|\mathbf{k}|)^2 \Phi(\mathbf{k})\Phi(-\mathbf{k})(\mathbf{k}^2 + m^2) \Big) - \frac{\hbar}{2} \int d\mathbf{k}\kappa(|\mathbf{k}|)\omega(k).$$

(6.49)

Then, the ground state is

$$\psi^\kappa = Z_\kappa^{-1} \exp\Big(-\frac{1}{2\hbar} \Phi \omega_\kappa \Phi \Big),$$

(6.50)

where

$$\omega_\kappa = \kappa\omega$$

(6.51)

and $\omega(k) = \sqrt{k^2 + m^2}$. We are going to solve the Schrödinger equation

$$i\hbar\partial_t \psi = H_\kappa \psi,$$

(6.52)

where $H_\kappa = H_0^\kappa + V$. First, we consider $V = 0$. We write $\psi = \psi_0^\kappa \chi$. Then, the equation for χ takes the form (6.43) with $\psi^g \to \psi^\kappa$. We consider an oscillatory process ($\sigma = \sqrt{\hbar i}$)

$$d\Phi_s^\kappa(\mathbf{k}) = -i\omega_\kappa(k)\Phi_s^\kappa(\mathbf{k})ds + \sigma dW_s(\mathbf{k}).$$

(6.53)

It has the solution (with the initial condition at t_0)

$$\Phi_s^\kappa(\mathbf{k}) = \exp(-i\omega_\kappa(k)(s - t_0))\Phi(\mathbf{k}) + \sigma \int_{t_0}^{s} \exp(-i\omega_\kappa(k)(s - \tau))dW_\tau(\mathbf{k}).$$

(6.54)

We calculate

$$\int d\Phi |\psi^\kappa(\Phi)|^2 E\Big[\exp(\int dt d\mathbf{x} f_t(\mathbf{x})\Phi_t^\kappa(\mathbf{x})) \Big]$$
$$= \exp\Big(\tfrac{1}{2} \int dt dt'\Big(f_t, (2\omega_\kappa)^{-1} \exp(-i\omega_\kappa|t - t'|)f_{t'} \Big) \Big).$$

(6.55)

In the calculation of (6.55) we assume that the measure $d\mu_\kappa = d\Phi |\psi^\kappa|^2$ is normalized . So that $\int d\mu_\kappa = 1$ and the integral over the intial condition in Eq. (6.54) is $\int d\mu_\kappa \Phi(\mathbf{x})\Phi(\mathbf{x}') = (2\omega_\kappa)^{-1}(\mathbf{x}, \mathbf{x}')$.

From Eqs. (6.54)–(6.55) the two-point function of time-ordered products of field operators is

$$i\Delta_F^\kappa(\mathbf{x}, \mathbf{x}') = \int d\mathbf{k} \exp(i\mathbf{k}(\mathbf{x} - \mathbf{x}'))(2\omega_\kappa)^{-1} \exp(-i\omega_\kappa|t - t'|).$$

If κ grows, e.g., as $\kappa = 1 + \epsilon \mathbf{k}^4$, then the field will be regular in $d < 5$ dimensions because

$$\left| i \triangle_F^\kappa(\mathbf{x}, \mathbf{x}') \right| \leq \int d\mathbf{k} (2\omega_\kappa)^{-1} < K_\kappa$$

for a certain constant K_κ.

In the limit $\kappa \to 1$ we obtain the generating functional for the time-ordered correlation functions of the quantum scalar free field Φ_t^Q

$$
\begin{aligned}
\lim_{\kappa \to 1} &\left(\psi^\kappa, T \left(\exp(\int dt \Phi_t^Q(f_t)) \right) \psi^\kappa \right) \\
&= \int d\Phi |\psi^g(\Phi)|^2 E \left[\exp(\int dt d\mathbf{x} f_t(\mathbf{x}) \Phi_t(\mathbf{x})) \right] \\
&= \exp \left(\tfrac{1}{2} \int dt dt' \left(f_t, (2\omega)^{-1} \exp(-i\omega |t - t'|) f_{t'} \right) \right).
\end{aligned}
\tag{6.56}
$$

Hence, vacuum correlation functions of time-ordered products of the quantum free field Φ^Q coincide with the stochastic expectation values. In the limit $t_0 \to -\infty$ we can drop the first term on the rhs of Eq. (6.54) as the oscillatory term appearing in the formula (6.54) is vanishing for $t_0 \to -\infty$ when calculating the expectation values (6.56) because of the Lebesgues lemma applied to

$$\int d\mathbf{k} \exp \left(- i(t - t_0)\kappa(k)\sqrt{\mathbf{k}^2 + m^2} \right) f(\mathbf{k}) \tag{6.57}$$

(when $|f(\mathbf{k})|$ is an integrable function). When $t_0 \to -\infty$ there is no need to average over ψ_0^κ in Eqs. (6.55) and (6.56).

In the Heisenberg picture we define the quantum field (where by $U_{(t,s)} = \exp(-iH(t - s))$ we denote the evolution from s to t and $U_{(t,0)} \equiv U_t$)

$$\Phi_t^Q = U_t^+ \Phi^Q U_t. \tag{6.58}$$

Then, we can derive the field expectation value in another way. We have for the two-point Wightman function

$$
\begin{aligned}
(\psi^g, \Phi_{-t'}^Q(\mathbf{x}') \Phi_{-t}^Q(\mathbf{x}) \psi^g) &= (U_{t'} \Phi^Q(\mathbf{x}') \psi^g, U_t \Phi^Q(\mathbf{x}) \psi^g) \\
&= Z^{-2} \int d\Phi \exp(-\Phi, \omega\Phi) E \left[\Phi_{t'}(\mathbf{x}') \right]^* E \left[\Phi_t(\mathbf{x}) \right] \\
&= \left((2\omega)^{-1} \exp(-i\omega(t - t')) \right)(\mathbf{x}, \mathbf{x}').
\end{aligned}
\tag{6.59}
$$

Still in another way for the time-ordered product for $t \geq t'$

$$Z^{-2} \int d\Phi \exp(-\Phi, \omega\Phi) E\left[\Phi_t(\mathbf{x})\Phi_{t'}(\mathbf{x}')\right]$$
$$= (\psi^g, T\left(\Phi_t^Q(\mathbf{x})\Phi_{t'}^Q(\mathbf{x}')\right)\psi^g) = (\psi^g, \Phi^Q(\mathbf{x})U_{t-t'}\Phi^Q(\mathbf{x}')\psi^g)$$
$$= Z^{-2} \int d\Phi \exp(-\Phi, \omega\Phi)\Phi(\mathbf{x}) E\left[\Phi_{t-t'}(\mathbf{x}')\right] \qquad (6.60)$$
$$= \left((2\omega)^{-1}\exp(-i\omega(t - t'))\right)(\mathbf{x}, \mathbf{x}').$$

This is so as ($t_0 = 0$)

$$E[\Phi_t(\mathbf{x})] = (\exp(-i\omega t)\Phi)(\mathbf{x}) \qquad (6.61)$$

and the integral with respect to the Gaussian measure $d\mu = Z^{-2}d\Phi \exp(-\Phi, \omega\Phi)$ is

$$\int d\mu \Phi(\mathbf{x})\Phi(\mathbf{x}') = (2\omega)^{-1}(\mathbf{x}, \mathbf{x}').$$

We add an interaction $V(\Phi)$ to the free Hamiltonian H_0^κ. The Feynman integral for the solution of the Schrödinger equation (6.52) with an initial condition $\psi = \psi^\kappa \chi$ reads

$$\psi_t^\kappa(\Phi) = \psi^\kappa E\left[\exp\left(-\frac{i}{\hbar}\int_{t_0}^t V(\Phi_s^\kappa)d\mathbf{x}ds\right)\chi\left(\Phi_t^\kappa(\Phi)\right)\right]. \qquad (6.62)$$

where Φ_t^κ is defined in Eq. (6.54). Then, it follows from the derivation of the Gell-Mann-Low formula that if we let in Eq. (6.54) $t_0 = -\infty$, define the field (neglecting the first term on the rhs of Eq. (6.54) according to the discussion at Eq. (6.57))

$$\Phi_t^\kappa = \sqrt{i\hbar}\int_{-\infty}^t \exp(-i\omega_\kappa(t - s))dW_s, \qquad (6.63)$$

then the correlation functions in the physical vacuum can be calculated according to the formula

$$Z^{-1}E\left[\exp\left(-\frac{i}{\hbar}\int_{-\infty}^\infty V(\Phi_s^\kappa)d\mathbf{x}ds\right)\Phi_{t_1}^\kappa \ldots \Phi_{t_n}^\kappa\right], \qquad (6.64)$$

where

$$Z = E\left[\exp\left(-\frac{i}{\hbar}\int_{-\infty}^\infty V(\Phi_s^\kappa)d\mathbf{x}ds\right)\right]. \qquad (6.65)$$

Equation (6.64) replaces the Gell-Mann-Low formula (3.37) which is expressed by vacuum expectation values in the Fock space of quantum fields Φ^Q

$$\Phi^Q = (2\pi)^{-\frac{3}{2}}\int d\mathbf{k}(2\omega)^{-1}(a(\mathbf{k})\exp(-ikx) + a(\mathbf{k})^+\exp(ikx))$$

defined in terms of the creation and annihilation operators as in Eq. (2.22). The Feynman formula in the Fock space has been obtained in ref. [85]. The benefit of Eqs. (6.62) and (6.64) is that they can be applied in numerical calculations of correlation functions (e.g., by means of the Monte Carlo methods).

If the interaction is of the exponential type as in Sect. 6.2

$$V(\Phi^\kappa(x)) = \int d\mu(a) \exp(ia\alpha\Phi^\kappa(x)) \tag{6.66}$$

then when expressed in the representation (6.54) we have

$$
\begin{aligned}
V(\Phi_s^\kappa(x)) = &\int d\mu(a) \\
&\times \exp\left(ia\alpha\left(\exp(-i\omega_\kappa(s - t_0))\Phi^\kappa(\mathbf{x}) + \sigma \int_{t_0}^s \exp(-i\omega_\kappa(s - \tau))dW_\tau\right)\right).
\end{aligned} \tag{6.67}
$$

Let us calculate (for illustration) the second order terms in the (Dyson) perturbation expansion of the solution (6.62) of the Schrödinger equation and a similar term in the correlation functions (6.64)–(6.65). The expectation value of the second order term in the perturbation expansion of $\exp(-\frac{i}{\hbar}\int_{t_0}^t ds V)$ is

$$
\begin{aligned}
\int d\mu(a_1) & d\mu(a_2) ds_1 ds_2 d\mathbf{x}_1 d\mathbf{x}_2 \\
&\times \exp\left(-\tfrac{1}{2}\alpha^2\sigma^2 E\left[\left(a_1 \int_{t_0}^{s_1} \exp(-i\omega_\kappa(s_1 - \tau))dW_\tau(\mathbf{x}_1)\right.\right.\right. \\
&\left.\left.\left. + a_2 \int_{t_0}^{s_2} \exp(-i\omega_\kappa(s_2 - \tau))dW_\tau(\mathbf{x}_2)\right)^2\right]\right)
\end{aligned} \tag{6.68}
$$

times the terms depending on initial conditions. The terms at the same $\mathbf{x}$ in Eq. (6.68) are removed by normal ordering. Let us consider the mixed term in the exponential which is

$$
\begin{aligned}
a_1 a_2 \alpha^2 \sigma^2 &\int d\mathbf{k} \exp(i\mathbf{k}(\mathbf{x}_1 - \mathbf{x}_2)) \int_{t_0}^{min(s_1,s_2)} \exp(-i\omega_\kappa(s_1 + s_2 - 2\tau))d\tau \\
= a_1 a_2 \alpha^2 \sigma^2 &\int d\mathbf{k} \exp(i\mathbf{k}(\mathbf{x}_1 - \mathbf{x}_2)) \\
&\times (2i\omega_\kappa)^{-1}\left(\exp(-i\omega_\kappa|s_1 - s_2|) - \exp(-i\omega_\kappa(s_1 + s_2 - 2t_0))\right).
\end{aligned} \tag{6.69}
$$

The calculations with the Gell-Mann-Low formula (6.64) are simpler because we can let $t_0 = -\infty$. Then, the terms depending on initial conditions are absent and the last term in Eq. (6.69) is vanishing. For the same reason because of the integration over $\mathbf{k}$ in $\omega(k)$ in Eq. (6.37) the term at $t_0 = -\infty$ is disappearing. Then, $|\exp F| \leq 1$ in Eq. (6.34). In such a case we can allow an infinite range of a (with regularized ω_κ) in $d\mu(a)$ if $\int d|\mu(a)| < \infty$. When $\omega_\kappa \to \omega$ then the integral (6.69) is singular at space-time coinciding points. In one spatial dimension discussed in the next section the singularity of (6.69) is logarithmic. In this case the exponential in Eq. (6.68) has power-law singularities which can be integrable for small a.

6.4 Feynman-Wiener Integration in QFT in Two Dimensions

The two-dimensional quantum field theory with a trigonometric interaction $\cos(a\Phi)$ (sine-Gordon model) is of interest (in the real time) for some conceptual developments in QFT related to the soliton classical solutions and charge sectors corresponding to various asymptotic conditions at $t \to \infty$ [94, 202]. A two-dimensional model with an exponential potential arises as a model of string theory (Polyakov model [199, 230]).

In one spatial dimension the correlation function in Eq. (6.69) is

$$a_1 a_2 \alpha^2 \sigma^2 \int_{-\infty}^{\infty} dk \exp(ik(x_1 - x_2))(2i\omega)^{-1}$$
$$\times \left(\exp(-i\omega|s_1 - s_2|) - \exp(-i\omega(s_1 + s_2 - 2t_0)) \right) \tag{6.70}$$

with $\omega = \sqrt{m^2 + k^2}$. It is easy to see that when $s_1 - s_2 \to 0$ then the correlation (6.70) behaves as

$$K a_1 a_2 \ln |x_1 - x_2|$$

for short distances $|x_1 - x_2|$ with a certain constant K (we have similar logarithmic behavior if $x_1 = x_2$ and $s_1 - s_2 \to 0$ and more generally if $(x_1 - x_2)^2 - (s_1 - s_2)^2 \to 0$). In such a case the second order term in Eq. (6.68) in the expansion of the solution of the Schrödinger equation (6.62) and in the Gell-Mann-Low formula (6.64) of Sect. 6.3 behaves as

$$\exp \left(K a_1 a_2 \ln |x_1 - x_2| \right) \simeq |x_1 - x_2|^{K a_1 a_2} \tag{6.71}$$

Hence, it has power-law singularities. For sufficiently small a_j the subsequent terms in the perturbation expansion are integrable. The expansion in a perturbation series of the Feynman-Wiener integral for the model (6.66) is similar to the one in the sine-Gordon model in Euclidean quantum field theory [63, 93] where the convergence of the perturbation series has been proved.

Next, let us consider the Polyakov model of an exponential interaction $V(\Phi)) = \int d\mu(a) \exp a\Phi$ (in the Polyakov model $d\mu(a)$ is concentrated at one point; for exponential interactions see [7, 11]) of a massless scalar field Φ [199, 230]. The two-point function of Φ is

$$D^{(+)}(x, t) = (2\pi)^{-1} \int_{-\infty}^{\infty} dp |p|^{-1} \exp(-i|p|t - ipx)$$
$$= (2\pi)^{-1} \left(\int_0^{\infty} dp\, p^{-1} \exp(-ip(t + x)) + \int_0^{\infty} dp\, p^{-1} \exp(-ip(t - x)) \right) \tag{6.72}$$
$$= (2\pi)^{-1} (\ln(x_+ - i0) + \ln(x_- - i0)),$$

where $x_+ = t + x$, $x_- = t - x$ and the notation $-i0$ means that this two-point function should be considered as a boundary value of a function analytic in the lower half-plane [214]. We consider

$$\int dx\, V(\Phi) = \int dx \int d\mu(a)\, \exp(a\Phi(t,x)). \qquad (6.73)$$

For the solution of the Schrödinger equation we can apply the perturbation expansion. The question is whether we can remove the regularization of Sect. 6.3 term by term. For this purpose we note that in the perturbation series at the Nth order we obtain (see Sect. 6.2)

$$\exp\left(\frac{1}{2}\sum_{j,k} a_j a_k E[\Phi(t_j, x_j)\Phi(t_k, x_k)]\right) \qquad (6.74)$$

which has power-law singularities according to Eq. (6.71) and the analysis of the previous section. These singularities are integrable for small a. The Feynman-Wiener integral shows that the perturbative expansions in sine-Gordon as well as the exponential models without ultraviolet cut-off can be treated in the real time framework instead of the imaginary time.

Summarizing this chapter we may say that working with the oscillatory paths of quantum fields we can achieve by means of the Feynman-Wiener integration the same perturbative results as for QFT in the Fock space. However, some formulas acquire a non-perturbative meaning. In these cases we can work with the Feynman integral (in the real time) as a rigorous tool in the same way as in Euclidean field theory [111, 124]. Some extensions for polynomial interactions beyond the class of models treated by the Euclidean approach are briefly discussed in Sect. 8.7.

6.5 Exercises

6.1 Differentiate the Dyson expansion (6.8) term by term and show that the sum is the solution of the Schrödinger equation.

6.2 Calculate the expectation value (6.13) and show that Eq. (6.12) is satisfied

6.3 Calculate the expectation value (6.55).
    Hint: $E[\exp \int f dw] = \exp(\frac{1}{2}E[(\int f dw)^2])$ if f does not depend on w. Then, use Eq. (5.56).

6.4 Prove Eq. (6.62) by explicit calculations for $V(\Phi) = \int d\mathbf{x} f(\mathbf{x})\Phi(\mathbf{x})$ using the Fourier reprepresentation

$$\chi(\Phi) = \int d\mu(\Sigma)\rho(\Sigma)\exp(i\int d\mathbf{x}\Sigma(\mathbf{x})\Phi(\mathbf{x})),$$

where μ is a Gaussian measure and ρ is an integrable function. This is reducing the check of Eq. (8.62) to $\chi = \exp(i\int d\mathbf{x}\Sigma(\mathbf{x})\Phi(\mathbf{x}))$ and Gaussian expectation values of exponentials of linear functions of the Wiener process.

Chapter 7
Application of the Feynman Integral for Approximate Calculations

Abstract In this chapter we discuss a semi-classical approximation in quantum mechanics and quantum field theory. We restrict ourselves to a formal expansion in $\hbar$ resulting in a direct way from the path integral. We formulate the semi-classical approximation as a stationary phase method of the calculation of integrals. We perform such an integral for harmonic and anharmonic oscillators. In application to QFT we show that the result of the stationary phase method can be considered as a resummation of the Feynman diagrams with one and more loops (depending on the power of the small parameter $\hbar$). In Euclidean field theory the corresponding method of computations can be associated with the Laplace method of an approximate calculation of integrals. The result of calculations by the Laplace method can be expressed as an effective action. The effective action in the first order of perturbation can be represented by a determinant of a differential operator. In this chapter the determinants are studied first in a perturbation expansion and next in an exact expression by means of the heat kernel using the Feynman-Kac path integral representation of the heat kernel.

In this chapter we discuss a semi-classical approximation in quantum mechanics and quantum field theory. There is a huge literature on this subject (especially concerning quantum mechanics [32]). The rigorous estimates are based either on the theory of differential equations [98] or on a rigorous version of path integrals [12, 13, 76, 138]. We restrict ourselves in this chapter to a formal expansion in $\hbar$ resulting in a direct way from the path integral. We return to the intuitive Feynman formulation of the "sum over paths". As explained in Sect. 5.1 the sum over paths has a mathematical justification in terms of the Trotter product formula together with the Riemannian approximation to the integral $\int V ds$ in the Feynman-Kac formula (5.38) giving finite-dimensional approximations (via the "cylinder function" integration (5.49)) to the infinite dimensional Feynman-Wiener integral. The formal calculations in this chapter could be understood as follows: we consider the finite dimensional (discrete) approximations to the Feynman integral, next we perform the stationary point approximation to finite dimensional integrals and subsequently we take the continuum limit of the discrete approximations (such an approach appears

Z. Haba, *Lectures on Quantum Field Theory and Functional Integration*,
https://doi.org/10.1007/978-3-031-30712-6_7

in [13]). Another mathematical approach to semi-classical estimates is based on a
rigorous version of the transformation of translation of the Gaussian measure (3.67)

$$
\begin{aligned}
d\mu_0(\phi + f) &= \exp\left(-\tfrac{1}{2\hbar}\int dx(\nabla_E f \nabla_E f + m^2 f^2)\right) \\
&\times \exp\left(-\tfrac{1}{\hbar}\int dx(\nabla_E \phi \nabla_E f + m^2 f\phi)\right)d\mu_0(\phi)
\end{aligned}
\tag{7.1}
$$

which follows from Eq. (3.66) (for the Wiener measure it is called the Cameron-
Martin theorem [108, 158]). In fact, this chapter could be developed on the basis of
this translation, where we choose f as a solution of classical equations. For detailed
estimates of the remainder in the expansion in $\hbar$ see refs. [76, 138, 224].

We formulate the semi-classical approximation as a stationary phase method of
the calculation of integrals in Sect. 7.1. In Sect. 7.2 we perform such an integral for
harmonic and anharmonic oscillators. In Sect. 7.3 it is shown that the result of the
stationary phase method can be considered as a resummation of the Feynman diagrams with one and more loops (depending on the power of the small parameter $\hbar$). In
Euclidean field theory the corresponding method of computations can be associated
with the Laplace method of an approximate calculation of integrals (Sect. 7.4). The
result of calculations is related to the notion of the effective action (Sect. 7.5). It can be
represented by a determinant of a differential operator.In Sect. 7.6 the determinants
are studied first in a perturbation expansion and next in an exact representation by
the heat kernel and the Feynman-Kac path integral representation of the heat kernel.

7.1 Semi-classical Expansion: The Stationary Phase Method

We explain the small parameter expansion of an oscillating integral taking as an
example the one dimensional integral of the form [89, 127, 154]

$$
. \; I = \int dx \exp\left(\frac{i}{\hbar}f(x)\right).
\tag{7.2}
$$

Let $f'(x_c) = 0$ at a certain point $x_c \in R$ then Taylor expansion around x_c has the
form

$$
f(x) = f(x_c) + \frac{1}{2}f''(x_c)(x - x_c)^2 + O((x - x_c)^3).
$$

Inserting this expansion in Eq. (7.2) and introducing a new variable $\sqrt{\hbar}y = (x - x_c)$
we obtain the integral (see Eq. (1.13))

$$
\begin{aligned}
I &= \sqrt{\hbar}\exp(\tfrac{i}{\hbar}f(x_c))\int dy \exp(\tfrac{i}{2}f''(x_c)y^2)\left(1 + O(\sqrt{\hbar})\right) \\
&= \sqrt{\hbar}\exp(\tfrac{i}{\hbar}f(x_c))\left(\tfrac{1}{2\pi}\left|f''(x_c)\right|\right)^{-\frac{1}{2}}\exp\left(\tfrac{i\pi}{4}\epsilon\left(f''(x_c)\right)\right).\left(1 + O(\sqrt{\hbar})\right),
\end{aligned}
\tag{7.3}
$$

where $\epsilon(x)$ is an antisymmetric function with $\epsilon(x) = 1$ when $x > 0$. There appears the square root of $f''(x_c)$ in the Gaussian integral (7.3). The sign of the square root is determined either by an analytic continuation as discussed at Eq. (1.13) (Sect. 1.3) or through the decomposition of both sides of the integral into the real and imaginary parts. If there are many points with $f'(x_c) = 0$ then we sum on the rhs of Eq. (7.3) over the contributions of these critical points. How the stationary phase integral looks like for paths contributing to the evolution kernel (5.16)? We have

$$\mathcal{L}(x(s)) = \mathcal{L}(x_c) + \frac{1}{2}\mathcal{L}''(x_c)(x(s) - x_c(s))^2 + O((x - x_c)^3), \qquad (7.4)$$

where the extremal path of the action satisfies the equation

$$m\frac{d^2}{ds^2}x_c(s) = -\nabla V(x_c(s)). \qquad (7.5)$$

This is the classical trajectory such that $x_c(0) = x$ and $x_c(t) = y$. We introduce a new variable $(q(0) = q(t) = 0)$

$$\sqrt{\hbar}q(s) = x(s) - x_c(s). \qquad (7.6)$$

Then, the stationary phase in the Feynman path integral reads (till $O(\sqrt{\hbar})$)

$$\left(\exp(-\tfrac{it}{\hbar}H)\right)(x, y) = \\ \sqrt{\hbar}\exp\left(\tfrac{i}{\hbar}\int_0^t \mathcal{L}(x_c(s))ds\right) \int \mathcal{D}q \exp\left(\tfrac{i}{2}\int_0^t ds q(s)\mathcal{L}''q(s)\right), \qquad (7.7)$$

where

$$\mathcal{L}'' = -\frac{d^2}{ds^2} - V''(x_c(s)) \qquad (7.8)$$

is the second order differential operator with boundary conditions $q(0) = q(t) = 0$. The q-integral is equal to (Eq. (1.13)) $\det(-\tfrac{i}{2\pi}\mathcal{L}'')^{-\frac{1}{2}}$. In the next section we apply Eq. (7.7) to quantum mechanics of an anharmonic oscillator.

7.2 Stationary Phase for an Anharmonic Oscillator

Let us begin with the harmonic oscillator. We would like to calculate the generating functional for the correlation functions of the harmonic oscillator

$$Z[u] = \int \mathcal{D}x(.) \exp\left(\frac{i}{\hbar} \int_{-\infty}^{\infty} ds(\mathcal{L}_0(x(s)) + u(s)x(s))\right). \qquad (7.9)$$

The action for the harmonic oscillator is ($m = 1$)

$$i \int_{-\infty}^{\infty} ds \mathcal{L}_0 = \frac{1}{2} \int_{-\infty}^{\infty} ds \left((\tfrac{dx}{ds})^2 - \omega^2 x^2 \right)$$

$$= \frac{i}{2} \int_{-\infty}^{\infty} dsx \left(- \tfrac{d^2}{ds^2} - \omega^2 \right) x \qquad (7.10)$$

$$= -\frac{1}{2} \int_{-\infty}^{\infty} xMx.$$

We have assumed here (when performing integration by parts) that at infinity $x = 0$. In Eq. (7.10)

$$M = i \left(\frac{d^2}{ds^2} + \omega^2 \right). \qquad (7.11)$$

Let

$$Jx = i \int ds u(s) x(s). \qquad (7.12)$$

We apply the formulas of Gaussian integration from Sect. 1.3

$$\int dX \exp\left(- X \tfrac{M}{2} X \right) \exp(JX) = \det \left(\tfrac{1}{2\pi} M \right)^{-\frac{1}{2}} \exp \left(- \tfrac{1}{2} J M^{-1} J \right)$$

and

$$\int dX \exp \left(- X \frac{M}{2} X \right) = \det \left(\frac{1}{2\pi} M \right)^{-\frac{1}{2}}. \qquad (7.13)$$

We do not need to care about the formula (7.13) because when the interval of s is $(-\infty, \infty)$ then this integral is just a constant (ω in M (7.11) can be used to rescale s). We must calculate $G = M^{-1}$, i.e., solve the equation

$$MG = 1. \qquad (7.14)$$

In terms of kernels (Sect. 3.4) Eq. (7.14) reads

$$i \left(\frac{d^2}{ds^2} + \omega^2 \right) G(s, s') = \delta(s - s') \qquad (7.15)$$

because the kernel of the unit operator is the δ-function. By direct differentiation (using $\partial_s \theta(s) = \delta(s)$) we check that

$$G(s, s') = \frac{1}{2\omega} \exp(-i\omega|s - s'|). \qquad (7.16)$$

Hence, in Eq. (7.13) we have

$$\int \mathcal{D}x(.)\exp\left(\frac{i}{\hbar}\int_{-\infty}^{\infty}ds\mathcal{L}_0\right)\exp\left(\frac{i}{\hbar}\int dsu(s)x(s)\right)$$
$$= \exp\left(-\frac{1}{2\hbar}\int u(s)u(s')G(s,s')\right). \qquad (7.17)$$

Expanding both sides of Eq. (7.17) in u (or by functional differentiation) we get

$$\int \mathcal{D}x(.)\exp\left(\frac{i}{\hbar}\int_{-\infty}^{\infty}ds\mathcal{L}_0\right)x(s)x(s') = \hbar G(s,s'). \qquad (7.18)$$

We know from quantum mechanics that the time-ordered correlation function of the position operator $x(s)$ in the Heisenberg picture in the ground state is

$$< 0|T(x(s)x(s'))|0 >= \hbar G(s,s'). \qquad (7.19)$$

Moreover, we have proved in Eq. (2.45) in QFT that

$$< 0|T\left(\exp\left(\frac{i}{\hbar}\int dsu(s)x(s)\right)\right)|0 >= \exp\left(-\frac{1}{2\hbar}\int u(s)u(s')G(s,s')\right). \qquad (7.20)$$

It follows that all time-ordered correlation functions of the harmonic oscillator position operator in quantum mechanics are the averaged values with respect to the Feynman integral calculated by means of the stationary phase method.

As a next problem we consider a computation of the evolution kernel of the harmonic oscillator via the path integral considered in Eq. (5.16) in Sect. 5.1. We are going to calculate the evolution kernel of the harmonic oscillator in an expansion in $\hbar$. The classical solution (the extremal path) is

$$x_c(s) = a\cos(\omega s) + b\sin(\omega s), \qquad (7.21)$$

where the constants a, b must be chosen so that $x_c(0) = x$ and $x_c(t) = y$. Expanding in Eqs. (7.4)–(7.7) $x(s) = x_c(s) + \sqrt{\hbar}q(s)$ where $q(0) = q(t) = 0$ we obtain

$$\left(\exp\left(-\frac{i}{\hbar}Ht\right)\right)(x, y) = \exp\left(\frac{i}{\hbar}\int_0^t \mathcal{L}_0(x_c(s))ds\right)\left(\frac{\hbar}{2\pi i}\det\left(-\frac{d^2}{ds^2}-\omega^2\right)\right)^{-\frac{1}{2}}, \qquad (7.22)$$

where the operator $-\frac{d^2}{ds^2}$ is defined on the interval $[0, t]$ with Dirichlet boundary conditions. The determinant in Eq. (7.22) is defined up to a normalization constant which is fixed by the requirement that at $t \to 0$ we have $(\exp(-\frac{i}{\hbar}Ht))(x, y) \to \delta(x - y)$. The eigenvalues of $-\frac{d^2}{ds^2}$ are $(\frac{\pi n}{t})^2$. In order to obtain the determinant we must calculate

$$\prod_{n=1}^{\infty}\left(\left(\frac{\pi n}{\omega t}\right)^2 - 1\right).$$

Such products are calculated in the theory of meromorphic functions (they are determined by the poles of the meromorphic function at $t = \frac{\pi n}{\omega}$ [208]). The result is

$$\left(\frac{\hbar}{2\pi i} \det\left(-\frac{d^2}{ds^2} - \omega^2\right)\right)^{-\frac{1}{2}} = \left(2\pi\hbar i \frac{\sin(\omega t)}{\omega}\right)^{-\frac{1}{2}}, \qquad (7.23)$$

where the factor $2\pi i$ comes from the above mentioned normalization to the δ-function. The $2\pi i$-factor could also be derived from the requirement that in the limit $\omega \to 0$ the evolution kernel must coincide with the one of Eq. (5.10). From Eqs. (7.22)–(7.23) we obtain the result known from quantum mechanics as the Mehler formula [187] (see [87]; Eq. (7.22) gives an exact formula because for quadratic Lagrangians there are no $O(\sqrt{\hbar})$ corrections).

What about the anharmonic oscillators? In quantum mechanics with $V = \frac{1}{2}\omega^2 x^2 + \tilde{V}$ (as follows from Sect. 3.3) the correlation functions in the physical ground state Φ_0 can be calculated as the correlation functions in the ground state $|0>$ of the harmonic oscillator via the Gell-Mann-Low formula

$$
\begin{aligned}
&< \Phi_0 | T(x(t_1)......x(t_n)) | \Phi_0 > \\
&= Z^{-1} < 0 | T\left(x(t_1)......x(t_n) \exp\left(-\frac{i}{\hbar}\int_{-\infty}^{\infty} ds\, \tilde{V}(x(s)))\right)\right) |0 > \qquad (7.24)\\
&= Z^{-1} \int \mathcal{D}x \exp\left(\frac{i}{\hbar}\int_{-\infty}^{\infty} ds\,(\mathcal{L}_0 - \tilde{V}(x(s)))\right) x(t_1)......x(t_n).
\end{aligned}
$$

Hence, we obtain the correlation functions by means of the path integral calculated with the Lagrangian $\mathcal{L} = \mathcal{L}_0 - \tilde{V}$. The independent perturbative proof of Eq. (7.24) follows from the results of Sects. 5.3 and 6.2 which were based on a comparison with the Dyson perturbative expansion in $\tilde{V}$.

We calculate the evolution kernel in another way through the $\hbar$ expansion applying the stationary point method to $\mathcal{L} = \mathcal{L}_0 - \tilde{V}$. We obtain instead of (7.22)

$$
\begin{aligned}
K_t(x, y) &\equiv (\exp(-\tfrac{i}{\hbar}H))(x, y) \\
&= \exp(\tfrac{i}{\hbar}\int_0^t ds\,\mathcal{L}(x_c(s)))\left(\tfrac{1}{2\pi i}\left(\det\left(-\tfrac{d^2}{ds^2} - \omega^2 - \tilde{V}''(x_c)\right)\right)\right)^{-\frac{1}{2}} \qquad (7.25)
\end{aligned}
$$

where x_c is the solution of

$$\frac{d^2}{ds^2}x_c + \omega^2 x_c = -\tilde{V}'(x_c) \qquad (7.26)$$

with the boundary conditions $x_c(0) = x$ and $x_c(t) = y$. There may be many solutions of this equation. Then, in Eq. (7.25) we must sum over these solutions. In quantum mechanics the determinant of the differential operator can be expressed by a finite dimensional determinant (van Vleck formula [32, 127, 184]). We must be careful in defining the square root of the determinant (see the discussion in Sect. 1.3) which depends on the path $x_c(s)$. The index in the square root has a geometric interpretation as the Morse-Maslov index of the path [127, 184]. The choice of phases in the square root in Eq. (7.25) is determined by the requirement that the evolution equation is to satisfy the composition law

$$\int dz K_t(x, z) K_s(z, y) = K_{t+s}(x, y) \tag{7.27}$$

Clearly, in order to get the kernel we must use approximate methods for the calculation of x_c. Nevertheless, the well-known (and very useful) WKB methods follow from the stationary phase method [32].

We have no Van Vleck formula in QFT. The determinants in QFT will be calculated either in perturbation theory or expressed by the solution of the heat equation as discussed in Sect. 7.6.

7.3 The Loop Expansion in QFT

We apply the stationary phase method to quantum field theory defined by the generating functional

$$Z[J] = \int d\phi \exp\left(\frac{i}{\hbar}\int dx(\mathcal{L}_0 - V + J\phi)\right) \tag{7.28}$$

where $\mathcal{L}_0$ is the Lagrangian of the free field. We calculate the generating functional according to the rules of Sect. 7.1 till an arbitrary order in $\hbar$. This expansion will resemble the Feynman diagram expansion, now with the propagator $(\partial_\mu \partial^\mu + m^2 + V''(\phi_c))^{-1}$ and vertices $\frac{1}{3!}V^{(3)}(\phi_c)\phi_q^3 + \frac{1}{4!}V^{(4)}(\phi_c)\phi_q^4 + \cdots$. From the Feynman rules of the perturbative expansion in Sect. 3.6 it can be seen that the number of loops L is related to the number of internal lines I and the number of vertices V by the formula (3.76) $L = I - V + 1$. On the other hand (from Sect. 7.1) the power P of $\hbar$ in the semi-classical expansion in $\hbar$ is

$$P = I - V. \tag{7.29}$$

Comparing Eqs. (3.76) and (7.29) we conclude that

$$P = L - 1. \tag{7.30}$$

Hence, the expansion in $\hbar$ is an expansion in the number of loops in the Feynman diagrams [56].

We can check these rules in the model

$$\mathcal{L} = \mathcal{L}_0 - V = \frac{1}{2}\partial_\mu\phi\partial^\mu\phi - \frac{1}{2}m^2\phi^2 - g\phi^4. \tag{7.31}$$

The stationary point ϕ_c is at

$$(\partial_\mu \partial^\mu + m^2)\phi_c + 4g\phi_c^3 = J. \tag{7.32}$$

Expanding (7.28) around ϕ_c, i.e., writing $\phi = \phi_c + \sqrt{\hbar}\phi_q$ we obtain

$$Z[J] = \sqrt{\hbar}\exp\left(\tfrac{i}{\hbar}\int \mathcal{L}(\phi_c)\right)\int d\phi_q \exp(-\tfrac{1}{2}\int \phi_q M\phi_q)(1 + O(\sqrt{\hbar}))$$
$$= \sqrt{\hbar}\exp\left(\tfrac{i}{\hbar}\int \mathcal{L}(\phi_c)\right)\det(\tfrac{M}{2\pi})^{-\tfrac{1}{2}}(1 + O(\sqrt{\hbar})), \tag{7.33}$$

where

$$M = i(\partial_\mu \partial^\mu + m^2 + 12g\phi_c^2). \tag{7.34}$$

We have calculated in Eq. (7.33) the generating functional till terms of order $\sqrt{\hbar}$. We could continue in this way the expansion in $\hbar$ (i.e., in the number of internal loops in the Feynman diagrams) till an arbitrary order with the propagator $(\partial_\mu \partial^\mu + m^2 + 12g\phi_c^2)^{-1}$. The result is non-perturbative in terms of the coupling constant g. We can show that if we expand (7.33) in the coupling constant then the formula (7.33) is just a resummation of Feynman diagrams with the fixed number of loops and an arbitrary number of vertices. For this purpose we write Eq. (7.32) as an integral equation

$$\phi_c = (\partial_\mu \partial^\mu + m^2)^{-1}(J - 4g\phi_c^3)$$
$$= -\Delta_F(J - 4g\phi_c^3). \tag{7.35}$$

We solve this equation by iteration. The lowest order is

$$\phi_c^{(0)} = -\Delta_F J. \tag{7.36}$$

The first order

$$\phi_c^{(1)} = 4g\Delta_F(\phi_c^{(0)})^3.$$

At any order in the expression (7.33) the result is represented by the Feynman propagators so that we can easily compare it with the standard perturbation expansion (Sect. 3.9). We recognize that $\exp(\tfrac{i}{\hbar}\int dx\mathcal{L}(\phi_c))$ gives the generating functional resulting from summing the Feynman diagrams with no loop (tree diagrams). The determinant in Eq. (7.33) comes from the summation of Feynman diagrams with one loop. In order to show this we write (with an infinite normalization constant Z)

$$Z^{-1}\det M = \det(1 - 12g\Delta_F\phi_c^2) \equiv \det(1 + gK) \tag{7.37}$$

and expand (7.37) in g as an expansion in powers of the integral operator K. This expansion will be discussed in Sect. 7.6.

7.4 The Saddle Point Method: The Loop Expansion in Euclidean Field Theory

In the Euclidean functional integral of Sects. 3.7–3.10 instead of the oscillatory integrals there appear exponentially decaying integrals. For simplicity we consider such an integral first in one dimension

$$I = \int dx \exp(-\frac{1}{\hbar} f(x)). \tag{7.38}$$

A derivation of the expansion in $\hbar$ of the integral (7.38) in a general case of a complex function f in the complex domain of $x \in C$ is called the saddle point method (or steepest descent). We restrict ourselves to a real function f defined on the real line with extremal points x_c on this line. This restricted steepest descent calculation is usually called the Laplace method.

Assume that $f'(x_c) = 0$ at certain $x_c \in R$ and $f''(x_c) > 0$. The Taylor expansion reads

$$f(x) = f(x_c) + \frac{1}{2} f''(x_c)(x - x_c)^2 + O((x - x_c)^3). \tag{7.39}$$

We change coordinates $x - x_c = \sqrt{\hbar} y$. Then, Eq. (7.38) gives

$$I(\hbar) = \sqrt{\hbar} \exp\left(-\frac{f(x_c)}{\hbar}\right)\left(\frac{1}{2\pi} f''(x_c)\right)^{-\frac{1}{2}}(1 + O(\sqrt{\hbar})). \tag{7.40}$$

If $f(x)$ has many minima then we sum over their contributions in Eq. (7.40).

We apply the Laplace method to the Euclidean path integral. According to Sect. 3.7 in Euclidean free field theory

$$< \exp(\phi(f)) > = \exp\left(\frac{1}{2} \int dx dy f(x) f(y) G_E(x - y)\right), \tag{7.41}$$

where the average is with respect to the functional measure

$$\begin{aligned} d\mu_0 &= d\phi \exp\left(-\frac{1}{2} \int dx (\nabla_E \phi \nabla_E \phi + m^2 \phi^2)\right) \\ &= d\phi \exp\left(-\frac{1}{2} \int dx \phi(-\triangle_E + m^2)\phi\right). \end{aligned} \tag{7.42}$$

Here ∇_E is the four-dimensional Euclidean gradient and $\triangle_E$ is the four-dimensional Euclidean Laplacian

$$\exp\left(\frac{1}{2} \int dx dy f(x) f(y) G_E(x - y)\right) = \int d\mu_0(\phi) \exp\left(\int dx f(x) \phi(x)\right). \tag{7.43}$$

Moreover, we can continue to the imaginary time also the Gell-Mann-Low formula
as follows

$$
\begin{array}{c}
< 0| T (\phi (x_1).....\phi (x_{2n}) \exp(-i \int V)) |0 > \to \\
\int d\mu_0 \exp(- \int dx V (\phi)) \phi (x_1).....\phi (x_{2n}).
\end{array}
\tag{7.44}
$$

We consider the model with the Euclidean Lagrangian

$$
\mathcal{L}_E = \mathcal{L}_0 + V = \frac{1}{2}\partial_\mu \phi \partial_\mu \phi + \frac{1}{2}m^2\phi^2 + g\phi^4.
\tag{7.45}
$$

We apply the Laplace method to the generating functional

$$
Z_E[J] = \int d\mu_0 \exp\left(- \frac{g}{\hbar} \int dx\phi^4 - \frac{1}{\hbar} \int dx J\phi \right) \simeq \int d\phi \exp\left(- \frac{1}{\hbar} \int (\mathcal{L}_E + J\phi)\right).
\tag{7.46}
$$

The stationary point ϕ_c is at

$$
(-\partial_\mu \partial_\mu + m^2)\phi_c + 4g\phi_c^3 = -J.
\tag{7.47}
$$

Expanding around ϕ_c, i.e., writing $\phi = \phi_c + \sqrt{\hbar}\phi_q$, we obtain

$$
\begin{aligned}
Z_E[J] &= \sqrt{\hbar} \exp\left(- \tfrac{1}{\hbar} \int \mathcal{L}_E(\phi_c)\right) \int d\phi_q \exp\left(- \tfrac{1}{2} \int \phi_q M \phi_q\right) \\
&= \sqrt{\hbar} \exp\left(- \tfrac{1}{\hbar} \int \mathcal{L}(\phi_c)\right) \det(\tfrac{1}{2\pi}M)^{-\frac{1}{2}},
\end{aligned}
\tag{7.48}
$$

where

$$
M = -\partial_\mu \partial_\mu + m^2 + 12g\phi_c^2.
\tag{7.49}
$$

The mathematical version of the derivation of the formula (7.48) relies on the trans-
lational transformation of the Gaussian measure (7.42) as discussed at the beginning
of this chapter

$$
\begin{aligned}
Z_E[J] &= \int d\mu_0(\sqrt{\hbar}\phi + \phi_c) \exp\left(- \tfrac{1}{\hbar} \int dx(g(\sqrt{\hbar}\phi + \phi_c)^4 + J(\sqrt{\hbar}\phi + \phi_c))\right) \\
&= \exp\left(- \tfrac{1}{\hbar} \int dx(\tfrac{1}{2}\nabla_E\phi_c\nabla_E\phi_c + \tfrac{1}{2}m^2\phi_c^2 + g\phi_c^4 + J\phi_c)\right) \int d\mu_0(\phi) \\
&\times \exp\left(- \tfrac{1}{\sqrt{\hbar}} \int dx(\nabla_E\phi\nabla_E\phi_c + m^2\phi\phi_c)\right) \exp\left(\tfrac{1}{\hbar} \int dx(-g(\sqrt{\hbar}\phi + \phi_c)^4 + g\phi_c^4 - \sqrt{\hbar}J\phi)\right)
\end{aligned}
$$

In this formula the term in the exponential linear in ϕ disappears if we choose ϕ_c as
the solution of Eq. (7.47). Then, performing the Gaussian integral of the remaining
quadratic term in the exponential we obtain Eq. (7.48). There are some substantial
difficulties if we wish to make this argument rigorous. First, the integral dx over an
infinite volume is not well-defined and should be treated as a limit of an expression
with a volume cutoff. Then, ϕ^4 needs renormalization. Hence, we must first make the
shift ϕ_c in a regularized theory and subsequently prove that the estimates survive the
removal of regularizations. The crucial point is that in the loop perturbation expansion
the products of propagators $(-\triangle_E + m^2 + 12g\phi_c^2)^{-1}$ require the same counterterms

as the ones $(-\triangle_E + m^2)^{-1}$ without the external field. The renormalization can be done perturbatively so that the renormalized saddle point result is just the sum of renormalized loop Feynman diagrams.

7.5 Effective Action

The effective action is a generalization of the notion of the classical potential energy when quantum fluctuations are taken into account. It can be considered as a result in QFT for the mean value of the quantum Hamiltonian calculated in a state with a mean value $\phi_c(x)$ of the quantum field [56].

Let us consider the generating functional in the form ($W(J)$ is the generating functional for connected Feynman diagrams)

$$Z[J] \equiv \exp(\frac{i}{\hbar} W(J)) = \int d\phi \exp\left(\frac{i}{\hbar} \int (\mathcal{L} + J\phi)\right). \tag{7.50}$$

Define the "classical field" as

$$\phi_c(x) = \frac{\delta W}{\delta J(x)}. \tag{7.51}$$

The effective action $\Gamma[\phi_c]$ is defined as the Legendre transform of W

$$\Gamma[\phi] = W[J] - \int dx J\phi_c. \tag{7.52}$$

It is understood that in Eq. (7.52) J is expressed by ϕ_c from Eq. (7.51) so that in Eq. (7.52) Γ is a functional expressed solely by ϕ_c. $\Gamma[\phi]$ can be expanded in a series

$$\Gamma[\phi] = \sum_n \frac{1}{n!} \int dx_1....dx_n \Gamma_n(x_1, ..., x_n)\phi(x_1), ..., \phi(x_n).$$

It can be shown that Γ_n gives n-point functions with amputated propagators of external lines (so called connected 1-particle irreducible diagrams).

Assume that we calculate W in an expansion in $\hbar$ [159]. We obtain till the first order

$$W = W_{cl}^{(0)}(\phi_0) + \hbar W_1(\phi_0), \tag{7.53}$$

where

$$W_{cl}^{(0)}(J) = \int dx \left(\frac{1}{2}\partial_\mu\phi\partial_\mu\phi - \frac{m^2}{2}\phi^2 - g\phi^4 + J\phi\right)$$

and from Eq. (7.32)

$$\partial_\mu\partial^\mu\phi_0 + m^2\phi_0 + 4g\phi_0^3 = J. \tag{7.54}$$

Then, from Eq. (7.51)

$$\phi_c = \frac{\delta W_{cl}^{(0)}}{\delta J(x)} + \hbar \phi_1. \tag{7.55}$$

For Γ till the first order in $\hbar$ we obtain from Eq. (7.52)

$$\Gamma(\phi_c) = W_{cl}^{(0)}(J) + \hbar W_1(\phi_0) - \int dx J \phi_c = \Gamma_{cl}^{(0)}(\phi_c) + \hbar W_1(\phi_c) + O(\hbar^{\frac{3}{2}}), \tag{7.56}$$

where

$$\Gamma_{cl}^{(0)}(\phi) = \int dx \left(\frac{1}{2} \partial_\mu \phi \partial_\mu \phi - \frac{m^2}{2} \phi^2 - g\phi^4 \right).$$

From our result on W_1 (Eq. (7.33) or Eq. (7.48) in Minkowski space-time)

$$\Gamma(\phi_c) = \Gamma_{cl}^{(0)}(\phi_c) + \frac{\hbar i}{2} \ln \det(\partial^\mu \partial_\mu + m^2 + 12 g \phi_c^2). \tag{7.57}$$

There are some minor changes in the Euclidean formulation for the effective action. So, Eq. (7.50) reads

$$Z_E[J] = \exp \left(-\frac{1}{\hbar} W(J) \right) = \int d\phi \exp \left(-\frac{1}{\hbar} \int (\mathcal{L}_E + J\phi) \right).$$

Equations (7.51)–(7.52) remain unchanged. Equation (7.57) takes the form

$$\Gamma_E(\phi_c) = \Gamma_{Ecl}^{(0)}(\phi_c) + \frac{\hbar}{2} \ln \det(-\triangle_E + m^2 + 12 g \phi_c^2), \tag{7.58}$$

where $\triangle_E$ is the four-dimensional Euclidean Laplacian and

$$\Gamma_{Ecl}^{(0)}(\phi) = \int dx \left(\frac{1}{2} \partial_\mu \phi \partial_\mu \phi + \frac{m^2}{2} \phi^2 + g\phi^4 \right).$$

7.6   Determinants of Differential Operators

At one loop of the effective action in $g\phi^4$ model we have the determinant (Eq. (7.37))

$$Z^{-1} \det M = \det(1 - 12g \triangle_F \phi_c^2) \equiv \det(1 + gK),$$

where Z is a (infinite) constant. First, let us calculate the rhs of Eq. (7.58) perturbatively using the formulas known for matrix determinants

$$\det(1 + gK) = \exp(Tr \ln(1 + gK)) = \exp\left(gTr(K) - \frac{1}{2}g^2 Tr(K^2) + \cdots\right).$$

$$(7.59)$$

The result is expressed by the Feynman propagators. It can be checked that the formula (7.59) agrees with standard perturbation expansion in QFT. For this purpose we calculate

$$gTr(K) = -12g \int dx \Delta_F(x, x) \phi_c^2(x). \qquad (7.60)$$

$\Delta_F(x, x)$ is quadratically divergent. We can identify the divergence (7.60) as the mass renormlization . Next

$$g^2 Tr(K^2) = \int dx dy K(x, y) K(y, x) = (12g)^2 \int dx dy \phi_c^2(x) \phi_c^2(y) (\Delta_F(x, y))^2.$$

$$(7.61)$$

Comparing with the coupling constant renormalization of Sect. 3.6 we can see that the integral (7.61) coincides with the Fourier transform of (3.40) calculated in Eq. (3.54) (the Feynman diagram (3.56)). Hence, the subtruction of the divergencies in Eq. (7.61) describes the coupling constant renormalization in the effective action (7.57). When we transform Eq. (7.61) to the momentum space then we can see that in order to make (7.57) finite we can apply the external momentum subtraction of Eq. (3.58). In coordinate space this means a subtraction of the term $\int dx \phi_c^4$ in Eq. (7.61). We can see that such a subtraction is renormalizing the term $g \int dx \phi_c^4$ in $\Gamma^{(0)}(\phi)$. We shall identify this counterterm in a non-perturbative renormalization of the one loop integrals by means of the heat kernel at the end of this section. $Tr(K^n)$ for $n \geq 3$ is already finite. There are no more divergencies in four-dimensions of the determinant (for more on the determinants, see [216, 222]).

A non-perturbative calculation of the determinant is based on the identities valid for determinants of matrices

$$\det M N^{-1} = \exp(Tr \ln M - Tr \ln N)$$
$$= \exp\left(-Tr \int_0^\infty dt[(M + t)^{-1} - (N + t)^{-1}]\right) \qquad (7.62)$$
$$= \exp\left(-Tr \int_0^\infty \frac{dt}{t}\left(\exp(-\frac{1}{2}Mt) - \exp(-\frac{1}{2}Nt)\right)\right).$$

Differential operators in the lattice approximation (Chap. 12) can be treated as matrices. This justifies the application of Eq. (7.62) for differential operators (for some other arguments see [217]).

We express the determinant (7.62) by a trace of the heat kernel $\exp(-\frac{1}{2}Mt)$ (see Eqs. (3.47)–(3.48)). The heat kernel of the second order differential operator can be conveniently represented by the path integral. Let us consider (Euclidean version $\mathbf{x} \in R^d$) the operator

$$M_u = -\Delta + 2u(\mathbf{x}) + m^2 \qquad (7.63)$$

and $N = M_0$. Then, as follows from Sect. 5.3

$$Tr(\exp(-tM_u)) = \int d\mathbf{x} \int dW_{(\mathbf{x},\mathbf{x})}^t(\mathbf{q}) \exp(-\frac{1}{2}m^2 t) \exp\left(-\int_0^t u(\mathbf{q}(s))ds\right),$$

$$(7.64)$$

where the path integration is over closed paths $\mathbf{q}(t) = \mathbf{q}(0) = \mathbf{x} \in R^d$. We can represent such paths as

$$\mathbf{q}(s) = \mathbf{x} + \sqrt{t}\mathbf{r}(\frac{s}{t}), \qquad (7.65)$$

where $\mathbf{r}(s)$ is the Brownian bridge. This is the Gaussian process defined on the interval $[0, 1]$ starting and ending in $\mathbf{0}$ with the covariance (see Sect. 5.3)

$$< r_j(s)r_k(s') > = \delta_{jk}\theta(s - s')s'(1 - s) + \delta_{jk}\theta(s' - s)s(1 - s'). \qquad (7.66)$$

There is a convenient representation of the Brownian bridge in terms of the Brownian motion $\mathbf{w}(s)$

$$\mathbf{r}(s) = (1 - s)\mathbf{w}\left(\frac{s}{1 - s}\right). \qquad (7.67)$$

In order to prove (7.67) (Exercise 7.4) it is sufficient to calculate the covariance of (7.67) and show that it coincides with (7.66) (Gaussian processes are determined by their covariance). We can now approach a calculation of the determinant using the expansion

$$u\left(\mathbf{x} + \sqrt{t}\mathbf{r}\left(\frac{s}{t}\right)\right) = u(\mathbf{x}) + \sqrt{t}\mathbf{r}\left(\frac{s}{t}\right)\nabla u + O(t). \qquad (7.68)$$

The measure corresponding to the Brownian bridge (7.67) in d-dimensions is related to $dW_{(\mathbf{x},\mathbf{x})}^t$ defined in Eq. (5.37)

$$\int dW_{(\mathbf{x},\mathbf{x})}^t(\mathbf{q})f(\mathbf{q}) = (2\pi t)^{-\frac{d}{2}} \int dW(\mathbf{w})f(\mathbf{q}(\mathbf{w})), \qquad (7.69)$$

where $dW(\mathbf{w})$ is the Wiener measure for paths starting from $\mathbf{x} = 0$ and $\mathbf{q}$ is expressed by Eqs. (7.65)–(7.67). With the representation (7.67) we have a relation between the integrals on both sides of Eq. (7.69). The equality (7.69) can be shown performing an integral with cylinder functions on both sides of this equation using the formula (5.37) on the lhs and the formula (5.32) for the Wiener integral on the rhs as discussed in Sect. 5.3.

The t-integral in Eq. (7.62) is

$$\int_0^\infty \frac{dt}{t}(Tr(\exp(-\frac{1}{2}tM_u) - \exp(-\frac{1}{2}tM_0))$$

$$= \int_0^\infty \frac{dt}{t} \int d\mathbf{x} \int dW_{(\mathbf{x},\mathbf{x})}^t(\mathbf{q}) \exp(-\frac{1}{2}m^2 t)\left(\exp(-\int_0^t u(\mathbf{q}(s))ds) - 1\right). \qquad (7.70)$$

In the integral (7.70) over t with the representation (7.65) we shall have (ultraviolet) divergencies at small t. The expansion of the exponential of the determinant (7.67) gives the proper time representation of the 1-loop Feynman diagrams of Sect. 3.8. The singularity in t in Eq. (7.70) is $t^{-\frac{d}{2}-1}$ (this can be considered as dimensional regularization discussed in Sect. 3.8 when we consider the determinant (7.62) in d-dimensions, it is also related to the ζ-function defnition of the determinant [148, 217]). We can expand the exponential in Eq. (7.70) in powers of t in order to detect the divergencies. In Eq. (7.70) the most divergent $t^{-\frac{d}{2}-1}$ term cancels. Then, from Eq. (7.68) there comes the $tu(\mathbf{x})$ term. In the effective action (with the interaction $g\phi^4$) $u = 6g\phi^2$. We expand (it is convenient to change the time integration variable $s = s't$ in order to follow the dependence of the integrals on t)

$$\exp(-\int_0^t u(\mathbf{q}(s))ds) - 1 = -\int_0^t (u(\mathbf{x}) + \sqrt{t}\,\mathbf{r}(\tfrac{s}{t})\nabla u)ds$$
$$+\tfrac{1}{2}\left(\int_0^t ds(u(\mathbf{x}) + \sqrt{t}\,\mathbf{r}(\tfrac{s}{t})\nabla u)\right)^2 + ... \tag{7.71}$$

The first divergent term in Eq. (7.70) is $\int dt\, t^{-\frac{d}{2}}u$. This term in the effective action is subtracted as the mass renormalization because $u \simeq \phi^2$. Then, there appear the $\sqrt{t}\,t^{-\frac{d}{2}-1}$ and $t\sqrt{t}\,t^{-\frac{d}{2}-1}$ terms multiplied by a derivative. Such terms after an integration over $\mathbf{x}$ are vanishing if there are no boundary contributions. If $d < 4$ then there will be no more divergencies. However, in $d = 4$ in the expansion of the exponential in powers of t we still encounter in Eq. (7.70) the logarithmic divergent integral $\frac{1}{2}\int dt\, t^{-1}u^2$ which in the ϕ^4 model gives the coupling constant renormalization. From Eqs. (7.70)–(7.71) the divergence of the effective action is

$$\frac{1}{4}(2\pi)^{-2} \int\limits_0^\infty \frac{dt}{t} \exp\left(-\frac{1}{2}m^2 t\right) \int (6g\phi_c^2)^2 dx. \tag{7.72}$$

After the subtraction of the divergent terms the effective action for large fields behaves as (see [135])

$$\Gamma(\phi) = g \int dx\phi^4 - g^2 C \int dx\phi^4 \ln\phi^2 + O(\partial\phi, \phi) \tag{7.73}$$

where $C > 0$ and O is bounded by ϕ^4. We leave the calculation of the constant C as an exercise (Exercise 7.7). We notice that at one loop the effective action in ϕ^4 theory is unbounded from below. The $-\phi^4 \ln\phi^2$ behavior (with the "wrong" sign) for large fields is characteristic to renormalizable field theories which are not asymptotically free [159]. The "correct" sign arises in gauge theories which are asymptotically free (see Sect. 11.6).

At one loop level it is easy to see that ϕ^6 in $d = 4$ is not renormalizable because in the expansion in t in Eq. (7.70) we have $u = 30g\phi^4$. Hence, the $\int dt\, t^{-\frac{d}{2}}$ counterterm is ϕ^4 and moreover we have the logarithmic divergent counterterm $\int \frac{dt}{t}\phi^8$. At one loop we can easily convince ourselves that the polynomials of the order $n > 4$ as well as non-polynomial interactions in $d = 4$ dimensions are not renormalizable because there will appear counterterms of the form V'' and $(V'')^2$ (confirming the power counting renormalizability result of Sect. 3.8).

7.7 The Functional Integral for Euclidean Fields at Finite Temperature

In Sect. 4.1 we calculated in the operator framework (in the Fock space) the correlation functions of the free scalar field in the thermal state. In this section we consider the interacting fields at an imaginary time till one loop in an expansion in $\hbar$. In a discussion of the continuation to an imaginary time the Kubo-Matrtin-Schwinger (KMS) condition [171, 183] plays an important role. For a Gibbs state KMS follows from the identity (for any observables A, B)

$$
\begin{aligned}
Tr\Big(A_t B \exp(-\beta H)\Big) &= Tr\Big(\exp(iHt - \beta H)A \exp(-iHt)B\Big) \\
&= Tr\Big(B \exp(iH(t + i\beta))A \exp(-iH(t + i\beta)) \exp(-\beta H)\Big) \\
&= Tr\Big(B A_{t+i\beta} \exp(-\beta H)\Big),
\end{aligned}
\tag{7.74}
$$

where $A_t = \exp(iHt)A \exp(-iHt)$.

From Eq. (7.74) it follows that if we continue in time as $t \to it$ then after the analytic continuation the expectation value of commuting observables A and B becomes periodic, i.e., invariant under the translation $t \to t + \beta$. This means that expressing the thermal correlation functions of quantum fields (after the analytic continuation to imaginary time) by a functional integral over Euclidean fields we integrate over fields which are periodic in time with the period β (see [121]). We can check this claim for free fields calculating the equal time correlation function by means of the functional integral with the use of the assumption of periodicity

$$
\begin{aligned}
Z^{-1} \int d\phi \exp\Big(-\tfrac{1}{2}\int dt d\mathbf{x}\phi(-\partial_t^2 - \Delta + m^2)\phi\Big)\phi(\tau, \mathbf{x})\phi(\tau, \mathbf{y}) \\
= (2\pi)^{-3} \int d\mathbf{p} \exp(i\mathbf{p}(\mathbf{x} - \mathbf{y})) \sum_{n=0}^{\infty}(n^2(\tfrac{2\pi}{\beta})^2 + m^2 + \mathbf{p}^2)^{-1} \\
= (2\pi)^{-3} \int d\mathbf{p} \exp(i\mathbf{p}(\mathbf{x} - \mathbf{y}))\tfrac{1}{2\omega(p)} \coth(\tfrac{1}{2}\beta\omega(p)).
\end{aligned}
\tag{7.75}
$$

The result coincides with Eq. (4.13) at $x_0 = y_0$ (Eucidean fields at $x_0 = y_0$ have the same correlation functions as quantum fields). In Eq. (7.75) we exploited the periodicity expanding the fields in the Fourier series

$$
\phi(t, \mathbf{x}) = \sum_{n=0}^{\infty} \exp\Big(i\frac{2\pi n}{\beta}t\Big)\Phi_n(\mathbf{x}).
\tag{7.76}
$$

Then, we applied the summation formula from the theory of meromorphic functions [208] (an expansion of a meromorphic function over its poles)

$$\sum_{n=1}^{\infty}(u^2 + n^2)^{-1} + \frac{1}{2u^2} = \frac{1}{2u}\pi \coth(\pi u) \qquad (7.77)$$

with $u = \frac{1}{2\pi}\beta\hbar\omega$.

The calculation of the effective action at one loop for the scalar fields with the interaction $V(\phi)$ by means of the functional integral in the Euclidean framework gives (see Eq. (7.58))

$$\Gamma(\phi_c) = \Gamma(\phi_c)^{(0)} + \frac{\hbar}{2}\ln \det \mathcal{G}$$

where

$$\mathcal{G} = -\frac{d^2}{dt^2} - \Delta + u \equiv -\frac{d^2}{dt^2} + \mathcal{M}^2, \qquad (7.78)$$

Δ is the three-dimensional Laplacian and $u = V''(\phi)$. The differential operator (in time) in Eq. (7.78) should be considered as a differential operator with periodic boundary conditions on the interval $[0, \beta]$ with eigenvalues $\frac{2\pi n}{\beta}$. We have

$$\det \mathcal{G} = \exp Tr \ln \mathcal{G} = -Tr \int_0^{\infty} dx \sum_{n=0}^{\infty}\left(\frac{4\pi^2}{\beta^2}n^2 + \mathcal{M}^2 + x\right)^{-1}$$

$$= -\frac{1}{2}Tr \ln \mathcal{M}^2 - \frac{\pi}{2}Tr \int dx \left(\frac{\beta^2}{4\pi^2}\mathcal{M}^2 + x\right)^{-\frac{1}{2}} \coth\left(\pi(\frac{\beta^2}{4\pi^2}\mathcal{M}^2 + x)^{\frac{1}{2}}\right) \quad (7.79)$$

$$= \frac{1}{2}Tr \ln \mathcal{M}^2 + Tr \ln\left(1 - \exp(-\beta|\mathcal{M}|)\right)$$

In the calculations in Eq. (7.79) we use Eq. (7.77) and we neglected some infinite constants (the integral (7.79) is divergent at the upper limit of x). The result is normalized so that when $\beta \to \infty$ then we obtain the formula (7.58) of Sect. 7.5 for the effective action at zero temperature in the Euclidean framework. The result agrees with the computations of [69]. We still discuss the electromagnetic and gravitational fields at finite temperature in Chaps. 9 and 10 (for more on fields at finite temperature, see [121, 168, 170]).

7.8 Exercises

7.1 Complete the computations (7.21)–(7.23) till the final Mehler formula for the evolution kernel.

7.2 In $g\phi^4$ model show (on the basis of Sect. 7.3) that the loop and a subsequent coupling constant expansion for n-point correlation functions coincide with the standard coupling constant expansion of Sects. 3.3 and 3.9.

7.3 Renormalize $Tr K^2$ of Eq. (7.59) in $g\phi^4$ model and show that this is the coupling constant renormalization of Sect. 3.6.

7.4 Calculate the covariance of $\mathbf{r}(s)$ Eq. (7.67) in order to prove that this is the Brownian bridge.

7.5 Compute the renormalization constant (7.72) in the effective action (7.70).

7.6 Prove Eq. (7.73) and compute the constant C in the effective action (7.73).

7.7 Perform the integral over x in Eq. (7.79) in order to get the final formula for the effective action.

Chapter 8
Feynman Path Integral in Terms of Expanding Paths

Abstract In the heuristic Feynman integral we integrate with respect to the Gaussian factor determined by the Lagrangian of a free particle. This is not a proper framework for quantum field theory where we expand around the free field (oscillator Lagrangian). In this chapter we further develop such an approach (initiated in Chap. 6) to functional integration with paths resulting from arbitrary Gaussian states. The Gaussian states are solutions of the Schrödinger equation for quadratic Hamiltonians. They are distinguished in quantum theory as the only states which have positively definite probability distribution in the phase space (the Wigner distribution). First, we discuss an expansion around paths of a particular solution of the Schrödinger equation. Then, we apply the method to QFT showing that paths in imaginary time can be identified with the Euclidean fields. The method of a construction of the stochastic process corresponding to a Gaussian solution of the Schrödinger equation is applied to quantum field theory on a manifold. We define the Schrödinger representation in Hilbert space for quantum fields on an expanding manifold on the basis of Gaussian solutions of the functional Schrödinger equation.

In the heuristic Feynman integral we integrate with respect to the Gaussian factor determined by the Lagrangian of a free particle. This is not a proper framework for quantum field theory where we expand around the free field (oscillator Lagrangian). We have already discussed in Chap. 6 a modification of the path integral so that the interaction perturbs the paths of an oscillator in its ground state instead of the paths of the free particles. In this chapter we further develop such an approach to functional integration with paths resulting from arbitrary Gaussian states. The Gaussian states are solutions of the Schrödinger equation for quadratic Hamiltonians. They satisfy the minimal uncertainty property but the uncertainties of position and momentum may be different. Hence, such states can be related to the squeezed states. For eigenstates of the time-independent Hamiltonians we may have (up to the trivial phase factor) time-independent solutions. If there are no normalized eigenstates (as for negative potentials) then there can still be time-dependent normalized solutions (the probability density $|\psi|^2$ is time-dependent). The time-dependent solutions are the only solutions for quantum theory of fields evolving on a space-time with a time-dependent metric (they can be Gaussian for a free field theory). The Gaussian states

© The Author(s), under exclusive license to Springer Nature Switzerland AG 2023
Z. Haba, *Lectures on Quantum Field Theory and Functional Integration*,
https://doi.org/10.1007/978-3-031-30712-6_8

are distinguished in quantum theory as the only states [157] which have positively definite probability distribution in the phase space (the Wigner distribution). From a mathematical point of view expanding around paths corresponding to Gaussian states ψ leads in QFT to the Hilbert space of states $L^2(d\mu)$ with the Gaussian functional measure $d\mu = |\psi|^2 d\phi$ which in contradistinction to the Lebesgues measure is well-defined in an infinite number of dimensions.

In Sect. 8.1 we discuss an expansion around paths of a particular solution of the Schrödinger equation. In Sect. 8.2 we choose as an example the upside-down oscillator. In Sect. 8.4 we study such expansions in the imaginary time. We apply the methods to QFT showing in Sect. 8.5 that paths in imaginary time can be identified with the Euclidean fields. The method of a construction of the stochastic process corresponding to a Gaussian solution of the Schrödinger equation is applied to quantum field theory on a manifold (Sects. 8.6–8.9). At the end of this chapter we discuss a typical problem in quantum mechanics with an application of the Feynman path integral: the interference of quantum waves. We show that with the stochastic representation it becomes an interference of waves with a stochastic phase.

8.1 Expansion Around a Particular Solution

In Chap. 6 we formulated the Feynman integral in terms of the paths of the oscillator in its ground state. In this chapter we are going to generalize the formulation representing the path integral in terms of paths corresponding to a time-dependent state. In Chap. 3 the correlation functions of interacting fields are expressed by correlation functions of free fields (Gell-Mann-Low formula).The formula applies to quantum mechanics: the correlation functions of the Heisenberg position operator of an anharmonic oscillator are expressed by correlation functions of the harmonic oscillator position operator. We wish to develop a formula to represent correlation functions in an arbitrary time-dependent state by a functional integral over paths determined by this state. If the spectrum of the Hamiltonian is continuous then the paths will not oscillate but rather expand.

First, we discuss the Schrödinger equation in one dimension when the potential of the harmonic oscillator is perturbed by a time-dependent potential V_t

$$i\hbar\partial_t\psi_t = \left(-\frac{\hbar^2}{2m}\nabla_x^2 + \frac{m\omega^2 x^2}{2} + V_t(x)\right)\psi_t \equiv H\psi_t. \tag{8.1}$$

We introduce a solution ψ_t^Γ of the Schrödinger equation for the oscillator potential (ω may depend on time)

$$i\hbar\partial_t\psi_t^\Gamma = \left(-\frac{\hbar^2}{2m}\nabla_x^2 + \frac{m\omega^2 x^2}{2}\right)\psi_t^\Gamma. \tag{8.2}$$

We represent the solution of Eq. (8.1) in the form

$$\psi_t = \psi_t^\Gamma \chi_t. \tag{8.3}$$

Then, χ_t solves the equation

$$\partial_t \chi_t = \left(\frac{i\hbar}{2m} \nabla_x^2 + \frac{i\hbar}{m} \nabla_x \ln \psi_t^\Gamma \nabla - \frac{i}{\hbar} V_t \right) \chi_t \tag{8.4}$$

with the initial condition $\chi_0 = \psi_0 (\psi_0^\Gamma)^{-1}$ determined by the initial conditions of ψ_t and ψ_t^Γ. With $V = 0$ the equation for χ is a diffusion equation with an imaginary diffusion constant and a complex drift. Its solution can be expressed by a solution of the Langevin equation (see Sect. 5.5)

$$dq_s = \frac{i\hbar}{m} \nabla \ln \psi_{t-s}^\Gamma (q_s) ds + \sqrt{\frac{i\hbar}{m}} dw_s, \tag{8.5}$$

The solution of Eq. (8.4) is given by the Feynman-Kac formula (see Sect. 5.5)

$$\chi_t(x) = E\left[\exp\left(-\frac{i}{\hbar} \int_0^t ds\, V_{t-s}(q_s(x)) \right) \chi_0(q_t(x)) \right], \tag{8.6}$$

here $q_s(x)$ is the solution of the Langevin equation (8.5) with the initial condition $q_0(x) = x$. The solution (8.6) has been discussed earlier in [138, 144] (for more mathematical details see [15, 71, 72, 137, 138, 141]). It is a real time version of the Feynman-Kac formula [96, 224].

We have proved the formula (8.6) in Sects. 5.2 and 5.5 for a class of analytic potentials and wave functions. The formula (8.6) can be proved directly by differentiation (using the Ito formula, see Sect. 5.4 and [158, 224]). For this purpose the composition law can be used $U(t, \tau) = U(t, s)U(s, \tau)$ which is equivalent to the Markov property (see [96]). In such a case in order to show that the formula (8.6) solves Eq. (8.4) it is sufficient to calculate the generator at $t = 0$. From Eq. (8.6) we have (when $t \to 0$)

$$d\chi_t = E[-\tfrac{i}{\hbar} V_0(x)\chi_0 + \nabla \chi_0 dq + \tfrac{1}{2} \nabla^2 \chi_0 dq dq].$$

We insert dq from Eq. (8.5), use $E[F dw_t] = 0$ and $dq dq = \frac{i\hbar}{m} dt$ (on the basis of the Ito calculus of Sect. 5.4). After the calculation of the differentials we may let $t \to 0$. Then, the rhs of $d\chi$ is $(\frac{i\hbar}{2m} \nabla^2 - \frac{i}{\hbar} V_0 + \frac{i\hbar}{m} \nabla \ln \psi_0^g \nabla) \chi dt$ (in agreement with the rhs of Eq. (8.4) at $t = 0$).

The solution of the stochastic equation (8.5) determines the correlation functions of the position operator x_t in the Heisenberg picture ($V = 0$)

$$(\psi_0^\Gamma, F_1(x_t) F_2(x) \psi_0^\Gamma) = \int dx |\psi_t^\Gamma(x)|^2 F_1(x) E\left[F_2\big(q_t(x)\big) \right]. \tag{8.7}$$

More general with $V \neq 0$

$$
\begin{aligned}
&(\chi\psi_0^\Gamma, F_1(x_t)F_2(x)\chi\psi_0^\Gamma) \\
&= \int dx |\psi_t^\Gamma(x)|^2 F_1(x) E\Big[\exp\Big(-\tfrac{i}{\hbar}\int_0^t ds V_{t-s}(q_s(x))\Big)\chi\Big(q_t(x)\Big)\Big]^* \\
&\times E\Big[\exp\Big(-\tfrac{i}{\hbar}\int_0^t ds V_{t-s}(q_s(x))\Big)F_2\Big(q_t(x)\Big)\chi\Big(q_t(x)\Big)\Big].
\end{aligned} \tag{8.8}
$$

8.2 The Upside-Down Oscillator

In this section we consider a quadratic potential (as in Eq. (8.2) with $\omega^2 = -\nu^2 < 0$) which is unbounded from below. The Schrödinger equation does not have normalized time-independent solutions. There are time-dependent solutions which do not belong to $L^2(dx)$. Nevertheless, they provide a drift to Eqs. (8.4)–(8.5). We discuss them at the end of this section. There are normalized time-dependent solutions of the form

$$
\psi_t^\Gamma = A(t)\exp\Big(-\frac{1}{2\hbar}x\Gamma_t x\Big). \tag{8.9}
$$

Equation (8.5) reads (we set $m = 1$ in this section)

$$
dq_s = -i\Gamma_{t-s}q_s ds + \sqrt{i\hbar}dw_s. \tag{8.10}
$$

Let $q_s(x)$ be the solution of Eq. (8.10) with the initial condition x. Then, the solution of the Schrödinger equation with the initial condition $\psi_0^\Gamma\chi$ is

$$
\psi_t = \psi_t^\Gamma E\Big[\chi\Big(q_t(x)\Big)\Big]. \tag{8.11}
$$

The model of an upside-down oscillator has been discussed by Guth and Pi [128] as an example of an exponential expansion (we preserve their notation). For a more traditional presentation see ref. [27]. The Schrödinger equation for this oscillator is

$$
i\hbar\partial_t\psi_t = \Big(-\frac{\hbar^2}{2}\nabla_x^2 - \frac{\nu^2 x^2}{2}\Big)\psi_t \equiv H\psi_t. \tag{8.12}
$$

The Gaussian solution of the Schrödinger equation reads

$$
\psi_t^\Gamma = A(t)\exp(-B(t)x^2) \tag{8.13}
$$

with

$$
A(t) = (2\pi)^{-\frac{1}{4}}\Big(\sqrt{\frac{\hbar}{\nu\sin 2\phi}}\cos(\phi - i\nu t)\Big)^{-\frac{1}{2}}
$$

and

$$B(t) = \tfrac{\nu}{2\hbar} \tan(\phi - i\nu t) = \tfrac{\nu}{2\hbar}$$
$$\times(\sin(2\phi) - i\sinh(2\nu t))(\cos(2\phi) + \cosh(2\nu t))^{-1}. \tag{8.14}$$

ϕ is expressing the initial condition for ψ_t^Γ as a Gaussian state with the variance $< x^2 > = (4B(0))^{-1} = (\tfrac{2\nu}{\hbar} \tan(\phi))^{-1}$. In the context of the imaginary time approach to quantum dynamics we note that when we continue (8.13) to an imaginary time $(t \to it)$ then $B(t)$ in Eq. (8.14) becomes negative for a sufficiently large t. Then, the wave function ψ_t^Γ (8.13) is growing exponentially fast at large x. Hence, the imaginary time approach discussed in Sects. 3.7, 5.3 and 8.4 is not applicable for the bottomless oscillator (8.12).

The stochastic equation (8.10) reads

$$dq_s = -2i\hbar B(t - s)q_s ds + \sqrt{i\hbar}dw_s.$$

Its solution is (it can grow exponentially in time)

$$q_s = \cos(\phi - i\nu(t - s))\Big(\cos(\phi - i\nu t)\Big)^{-1} x$$
$$+\sqrt{i\hbar}\cos(\phi - i\nu(t - s))\int_0^s \Big(\cos(\phi - i\nu(t - \tau))\Big)^{-1} dw_\tau. \tag{8.15}$$

We can calculate the evolution of the state $\chi(x)\exp(-B(0)x^2)$ using Eq. (8.11).

Let us consider now the upside-down oscillator in an approach of Chap. 6 with $\omega^2 = -\nu^2 < 0$. An analog of Eq. (6.27) corresponds to the Ansatz ($\nu > 0$)

$$\psi_t(x) = A(t)\exp\Big(\pm i\frac{\nu x^2}{2\hbar}\Big)\chi_t(x), \tag{8.16}$$

where $\exp(\pm i\frac{\nu x^2}{2\hbar})$ is an (unnormalized) eigenstate of the Hamiltonian (8.12) with an imaginary eigenvalue $\mp\frac{\nu}{2}i\hbar$ ($A(t) = \exp(\mp\frac{\nu}{2}t)$). The stochastic equation (6.31) takes the form ($\sigma = \sqrt{i\hbar}$)

$$dq_s = \mp\nu q_s ds + \sigma dw_s \tag{8.17}$$

When $\sigma \to 0$ then Eq. (8.17) is just the semiclassical Jacobi equation $\frac{dq}{dt} = \nabla W$ where W is the classical action (a solution of the Hamilton-Jacobi equation). Equation (8.17) has the solution (expanding paths for $\mp\nu > 0$)

$$q_t(x) = \exp(\mp\nu t)x + \sigma \int_0^t \exp(\mp\nu(t - s))dw_s \tag{8.18}$$

For the class of potentials discussed at the end of Sect. 6.1 (Eq. (6.3)) we can write the Feynman-Kac formula (for the solution of Eq. (8.1) with $\omega^2 = -\nu^2$) in the form (5.39)

$$\psi_t(a\sigma x) = \exp\left(\mp\frac{\nu}{2}t\right)\exp\left(\pm i\frac{\nu x^2}{2\hbar}\right)E\left[\exp\left(-\frac{i}{\hbar}\int_0^t V(q_s(a\sigma x))ds\right)\chi(q_t(a\sigma x))\right] \quad (8.19)$$

which applies if $E[|\chi(q_t(a\sigma x))|] < \infty$. Eq. (8.19) may have useful applications to the double well potentials $U = -\nu^2 x^2 + gx^4$ for a study of the tunnelling phenomena [165]. In Sect. 8.7 we discuss the corresponding Feynman-Kac formula in QFT.

It is instructive (Exercise 8.5) to check the formula (8.19) for $V = 0$ or $V = \alpha x$ for the Gaussian wave function (8.9) with the initial condition $\Gamma_0 = \mp i\nu x^2 + \gamma x^2$ ($\gamma > 0$). Then, the solution of Schrödinger equation (8.12) is in the form (8.9) (see Sect. 8.7) whereas the Gaussian expectation value in Eq. (8.19) can easily be calculated.

8.3 Solution in the Heisenberg Picture

In quantum mechanics the position operator is the counterpart of the field operator. In the Heisenberg picture

$$x_t = U_t^+ x U_t. \quad (8.20)$$

The model (8.12) is soluble in the Heisenberg picture

$$x_t = x\cosh(\nu t) + \frac{\sinh \nu t}{\nu}p. \quad (8.21)$$

One can easily calculate the expectation values (8.7) and (8.19) ($V = 0$) using Eq. (8.21) and check that the stochastic formula and the formula in the Heisenberg picture coincide. The solution of the Heisenberg equations of motion may serve as a supplementary tool in addition to solutions of Eq. (8.5). However, the functional integration formulas (8.6) with a potential may go beyond the perturbative calculations. For general potentials and functions of q it may be easier to calculate functional integrals (especially numerically) instead of correlation functions of the creation and annihilation operators.

Assume that we have the potential

$$U = -\frac{\nu^2}{2}x^2 + V(x) \quad (8.22)$$

such that the Hamiltonian has a ground state ($V(x)$ grows to $+\infty$ faster than quadratically). Then, there is a ground state for this system but there is no Gell-Mann-Low formula for the potential (8.22) because there is no ground state for the potential (8.12). Nevertheless, we can calculate the correlation functions which follow from the Feynman formula (8.6) and Eq. (8.7)

$$(\psi_0^\Gamma, x_t x \psi_0^\Gamma) = \int dx |\psi_t^\Gamma(x)|^2$$
$$\times E\left[\left(\exp(-i\int_0^t V(q_{t-s}(x))ds)\right)\right]^* x E\left[q_t(x)\exp\left(-i\int_0^t V(q_{t-s}(x))ds\right)\right].$$

$$(8.23)$$

A calculation of the expectation value in Eq. (8.23) can easily be achieved if we can calculate the expectation values of q_t. This is an interesting problem because a non-perturbative calculation with a polynomial V would describe the tunnelling between the minima of the potential U (8.22) [165].

As an alternative to the functional integration method for the potential (8.22) we consider the Heisenberg equations of motion

$$\left(-\frac{d^2}{dt^2} + \nu^2\right)x_t = -V'(x_t).$$

Using the Green function for the operator on the lhs

$$G(t,s) = \frac{1}{2\nu}\exp(-\nu|t-s|)$$

we rewrite the Heisenberg equation of motion as an integral equation

$$x_t = x\cosh(\nu t) + \frac{\sinh \nu t}{\nu}p - \int_0^t G(t,s)V'(x_s)ds \qquad (8.24)$$

which can be solved by iteration. Eq. (8.24) supplies a simple method to calculate perturbatively correlation functions. However, for non-perturbative estimates and numerical calculations the stochastic formula (8.23) may be more efficient. We apply the Heisenberg equations in Chaps. 9 and 10 for a description of motion of a quantum particle in quantum electromagnetic and gravitational fields.

8.4 Quantum Mechanics at an Imaginary Time

We return to the perturbations of the harmonic potential of the form (8.1) with $\omega^2 > 0$ at an imaginary time. Such models are well-explored in quantum mechanics and quantum field theory [9, 10, 78, 165]. We begin with the imaginary time version of the harmonic oscillator (Eq. (6.31)). The imaginary time stochastic equation is

$$dq_t = -\omega q_t dt + \sqrt{\frac{\hbar}{m}}dw_t. \qquad (8.25)$$

The solution with the initial condition x at t_0 reads

$$q_t = \exp(-\omega(t-t_0))x + \sqrt{\frac{\hbar}{m}}\int_{t_0}^t \exp(-\omega(t-s))dw_s. \qquad (8.26)$$

This is the Ornstein-Uhlenbeck process [52]. We may let $t_0 \to -\infty$ then the first term on the rhs of Eq. (8.26) disappears and we obtain a simple formula for correlation functions

$$E[q_t q_s] = \frac{1}{2\omega} \exp(-\omega |t - s|).$$

We can express the analytic continuation in time of the time-ordered correlation functions of the Heisenberg position operator in the ground state Φ_0 ($H \Phi_0 = 0$) in the theory with the Hamiltonian

$$H = H_0 + V,$$

where H_0 is the oscillator Hamiltonian. The result is [58, 221]

$$< q(t_1)......q(t_n) >= Z^{-1} E\left[\exp\left(-\int\limits_{-\infty}^{\infty} V(q(s))ds \right) q(t_1)......q(t_n)\right] \quad (8.27)$$

with

$$Z = E\left[\exp\left(-\int\limits_{-\infty}^{\infty} V(q(s))ds \right)\right].$$

This is the imaginary time Gell-Mann-Low formula. In fact, the formal derivation of the formula of Eq. (3.37) with an application of the oscillatory integrals has a rigorous imaginary time version [58, 111, 221] when the convergence of the Fock vacuum to the physical vacuum in the imaginary time can be rigorously established. In Eq. (8.27) the indefinite integral $\int_{-\infty}^{\infty} V$ must be understood as a limit which exists for potentials V such that there is the unique ground state for the Hamiltonian H.

In subsequent sections we discuss expectation values of quantum fields in Gaussian states which are not the ground states of H_0. In quantum mechanics we consider the imaginary time Schrödinger equation with the time-dependent potential V_t and (possibly) time-dependent frequency ω (a repetition of Eqs. (8.1)–(8.6) for an imaginary time)

$$- \hbar \partial_t \psi_t = \left(-\frac{\hbar^2}{2m} \nabla_x^2 + \frac{m\omega^2 x^2}{2} + V_t(x) \right) \psi_t \equiv H \psi_t. \quad (8.28)$$

First, let us look for solutions of the imaginary time Schrödinger equation for the harmonic oscillator

$$- \hbar \partial_t \psi_t^\Gamma = \left(-\frac{\hbar^2}{2m} \nabla_x^2 + \frac{m\omega^2 x^2}{2} \right) \psi_t^\Gamma. \quad (8.29)$$

We introduce a Gaussian solution of Eq. (8.29) (ω may depend on time)

$$\psi_t^\Gamma(x) = N(t) \exp\left(-\frac{1}{2\hbar}\Gamma_t x^2\right). \tag{8.30}$$

If we define $u(t) = \exp(\frac{1}{m}\int_0^t \Gamma_s ds)$ then u satisfies a linear equation

$$\partial_t^2 u - \omega^2 u = 0. \tag{8.31}$$

From $u(t)$ we can obtain $\Gamma(t) = mu(t)^{-1}\frac{du}{dt}$.

We represent the solution of Eq. (8.28) in the form

$$\psi_t = \psi_t^\Gamma \chi_t. \tag{8.32}$$

Then, χ_t solves the equation

$$\partial_t \chi_t = \left(\frac{\hbar}{2m}\nabla_x^2 + \frac{\hbar}{m}\nabla_x \ln \psi_t^\Gamma \nabla - \frac{1}{\hbar}V_t\right)\chi_t \tag{8.33}$$

with the initial condition $\chi_0 = \psi_0(\psi_0^\Gamma)^{-1}$. The solution of Eq. (8.33) can be expressed by the solution of the Langevin equation [96]

$$dq_s = \frac{\hbar}{m}\nabla \ln \psi_{t-s}^\Gamma(q_s)ds + \sqrt{\frac{\hbar}{m}}dw_s = -\Gamma_{t-s}q_s ds + \sqrt{\frac{\hbar}{m}}dw_s. \tag{8.34}$$

The solution of Eq. (8.28) is obtained by means of the Feynman-Kac formula [96, 224] (where the average $E[\ldots]$ is over the Brownian motion)

$$\chi_t(x) = E\left[\exp\left(-\frac{1}{\hbar}\int_0^t ds V_{t-s}(q_s(x))\right)\chi_0(q_t(x))\right]. \tag{8.35}$$

Here $q_s(x)$ is the solution of the Langevin equation (8.34) with the initial condition $q_0(x) = x$. The correlation functions in the state ψ_0^Γ are calculated as

$$(\exp(-tH)\psi_0^\Gamma, x \exp(-tH)(x\psi_0^\Gamma))$$
$$= \int dx |\psi_t^\Gamma|^2 x E\left[q_t(x)\exp\left(-\frac{1}{\hbar}\int_0^t ds V_{t-s}(q_s(x))\right)\right].$$

8.5 Paths at Imaginary Time as Euclidean Fields

We can follow the scheme of Sect. 8.4 in an infinite dimensional setting [10, 60, 122, 172]. Let ψ_t^g be a solution (usually the ground state) of the imaginary time Schrödinger equation

$$-\hbar\partial_t\psi = H\psi, \tag{8.36}$$

where H is given by Eq. (3.2). Let us consider the general solution of the Schrödinger equation (8.36) with the initial condition

$$\psi = \psi_0^\Gamma \chi. \tag{8.37}$$

Then, χ satisfies the equation

$$\hbar \partial_t \chi = - \int d\mathbf{x} \Big(\frac{1}{2} \Pi^2 + (\Pi \ln \psi_t^\Gamma) \Pi \Big) \chi, \tag{8.38}$$

where

$$\Pi(\mathbf{x}) = -i\hbar \frac{\delta}{\delta \Phi(\mathbf{x})}. \tag{8.39}$$

Equation (8.38) is a diffusion equation in infinite dimensional space [10, 122] of functionals belonging to $L^2(d\mu_t)$, where $d\mu_t = |\psi_t^\Gamma|^2 d\Phi$. Then, the solution of Eq. (8.38) can be expressed as

$$\chi_t(\Phi) = E\Big[\chi\big(\Phi_t(\Phi)\big)\Big], \tag{8.40}$$

where $\Phi_t(\Phi)$ is the solution of the stochastic equation

$$d\Phi_s(\mathbf{x}) = \hbar \frac{\delta}{\delta \Phi(\mathbf{x})} \ln \psi_{t-s}^\Gamma dt + \sqrt{\hbar} dW_s(\mathbf{x}) \tag{8.41}$$

with the initial condition Φ. $E[...]$ denotes an expectation value with respect to the Wiener process (Brownian motion) defined by the covariance

$$E\Big[W_t(\mathbf{x})W_s(\mathbf{y})\Big] = min(t,s)\delta(\mathbf{x} - \mathbf{y}). \tag{8.42}$$

Let us consider the simplest example: the free field. Then, a time-independent solution (the ground state) is

$$\psi^g = Z^{-1} \exp\Big(-\frac{1}{2\hbar} \Phi \omega \Phi \Big), \tag{8.43}$$

where Z is the normalization constant such that $\int d\mu_0 = 1$ where $d\mu_0 = |\psi^g|^2 d\Phi$. The stochastic equation (8.41) reads (this is the the imaginary time version of Eq. (6.53))

$$d\Phi_t = -\omega \Phi_t dt + \sqrt{\hbar} dW_t. \tag{8.44}$$

This is a well-defined stochastic equation in $L^2(d\mu_0)$, where $d\mu_0$ is the Gaussian measure with the covariance $(2\omega)^{-1}$. The solution is

$$\Phi_t(\Phi) = \exp(-\omega(t - t_0))\Phi + \sqrt{\hbar} \int_{t_0}^t \exp(-\omega(t-s))dW_s. \tag{8.45}$$

We calculate

$$E\left[\exp(\int dt d\mathbf{x} f_t(\mathbf{x}) \Phi_t(\Phi, \mathbf{x}))\right] d\mu_0(\Phi)$$
$$= \exp\left(\tfrac{1}{2} \int dt dt' \left(f_t, (2\omega)^{-1} \exp(-\omega|t - t'|) f_{t'}\right)\right). \tag{8.46}$$

If $t_0 = -\infty$ then $\Phi_t(\Phi)$ does not depend on the initial condition Φ and the integration over Φ in Eq. (8.46) is not required. We recognize the rhs of Eq. (8.46) as the generating functional for the Euclidean free field (3.87).

The imaginary time evolution generated by the Hamiltonian (3.2) with an interaction $V(\Phi)$ can be derived as in Eq. (8.35)

$$\chi_t(\Phi) = E\left[\exp\left(-\int_{t_0}^{t} ds d\mathbf{x} V(\Phi_s(\Phi))\right) \chi\left(\Phi_t(\Phi)\right)\right]$$

with the regularization of Sect. 6.3.

8.6 Free Field on a Static Manifold

We return to the real time evolution and we apply the framework of Sect. 8.1 to QFT on a manifold. This leads to a formulation of the scheme (6.26)–(6.33) in application to a quantum field defined on a Riemannian manifold discussed in an operator framework in Sect. 4.2. We consider the model of Sect. 4.2 of a quantum field on a static manifold with the heuristic Hamiltonian

$$H = \int d\mathbf{x} T_{00}(\mathbf{x}) = \int d\mathbf{x} \left(-\frac{\hbar^2}{2} g_{00} \frac{1}{\sqrt{g}} \frac{\delta^2}{\delta\phi(\mathbf{x})^2} + \tfrac{1}{2}\sqrt{g} g^{jk} \partial_j \phi \partial_k \phi + \frac{m^2}{2}\sqrt{g}\phi^2\right). \tag{8.47}$$

We write the wave functions in the form

$$\psi_t = \psi^g \chi_t,$$

where the ground state $\psi^g = \exp(-\frac{1}{2\hbar}(\phi, K\phi))$ has been derived in Sect. 4.2 (Eq. (4.27)). From the Schrödinger equation for ψ we obtain an equation for χ

$$\partial_t \chi = i\hat{H}\chi = \left(i\hbar \int d\mathbf{x} \frac{1}{2} g_{00} \frac{1}{\sqrt{g}} \frac{\delta^2}{\delta\phi(\mathbf{x})^2} - i \int d\mathbf{x} (K\phi)(\mathbf{x}) \frac{\delta}{\delta\phi(\mathbf{x})}\right)\chi. \tag{8.48}$$

Here, $\hat{H} = (\psi^g)^{-1} H \psi^g$ is a well-defined operator in the Hilbert space $L^2(d\mu_0)$ with the Gaussian measure μ_0 defined in Sect. 4.2, Eq. (4.31). We can express the solution of Eq. (8.48) in terms of the solution of the stochastic equation

$$d\phi_t(\mathbf{x}) = -i(K\phi_t)(\mathbf{x})dt + \sqrt{i\hbar}\sqrt{g_{00}}g^{-\frac{1}{4}}dW_t. \tag{8.49}$$

The solution of Eq. (8.49) is (we write explicitly the dependence of the solution on the initial condition ϕ)

$$\phi_t(\phi, \mathbf{x}) = \exp(-iK(t - t_0))\phi(\mathbf{x}) + \sqrt{i\hbar} \int_{t_0}^{t} \exp(-iK(t - s))\sqrt{g_{00}}g^{-\frac{1}{4}}dW_s.$$

The solution has the correlation functions

$$\int d\mu_0(\phi) E\Big[\phi_t(\phi, \mathbf{x})\phi_{t'}(\phi, \mathbf{x'})\Big]$$
$$= \lim_{t_0 \to -\infty} E\Big[\phi_t(\phi, \mathbf{x})\phi_{t'}(\phi, \mathbf{x'})\Big] = (\psi^g, T(\phi_t^Q(\mathbf{x})\phi_{t'}^Q(\mathbf{x'}))\psi^g) \tag{8.50}$$
$$= \hbar\sqrt{g_{00}(\mathbf{x})g_{00}(\mathbf{x'})}g^{-\frac{1}{4}}(\mathbf{x})g^{-\frac{1}{4}}(\mathbf{x'})\Big((2K)^{\hat{a}^{.}1} \exp(-iK|t - t'|)\Big)(\mathbf{x}, \mathbf{x'}),$$

where ϕ_t^Q is the quantum field. We could obtain this result also from Eq. (6.60) (now $\omega \to K$)

$$(\psi^g, \phi_t^Q(\mathbf{x})\phi^Q(\mathbf{x'})\psi^g) = \int d\phi|\psi^g|^2\phi(\mathbf{x})E[\phi_t(\phi, \mathbf{x'})]$$
$$= \hbar\sqrt{g_{00}(\mathbf{x})g_{00}(\mathbf{x'})}g^{-\frac{1}{4}}(\mathbf{x})g^{-\frac{1}{4}}(\mathbf{x'})\Big((2K)^{-1} \exp(-iKt)\Big)(\mathbf{x}, \mathbf{x'}).$$

The difference between (8.50) and (4.35) comes from the definition of the kernel in $L^2(d\nu)$, where ν is defined in Eq. (4.23), instead in $L^2(d\mathbf{x})$ (as explained in Sect. 3.4, Eq. (3.50)). The result (8.50) means that the field $\phi_t^Q(\mathbf{x})$ $= \sqrt{g_{00}}(\mathbf{x})g^{-\frac{1}{4}}(\mathbf{x})\phi_t(\mathbf{x})$ where ϕ on the rhs is defined in Eq. (4.33) by creation and annihilation operators in the Fock space.

The complex field ϕ (8.49) satisfies the randomly perturbed wave equation. We can decompose $\phi = \phi_1 + i\phi_2$ into real components. Let us introduce $\phi_+ = \phi_1 + \phi_2, \phi_- = \phi_1 - \phi_2$. Then, Eq. (8.49) is equivalent to the equations $\partial_t\phi_- = K\phi_+$ and

$$\partial_t^2\phi_- = -K^2\phi_- + K\sqrt{2\hbar}\sqrt{g_{00}}g^{-\frac{1}{4}}\partial_t W. \tag{8.51}$$

This is an analog of Eq. (6.32) derived for the quantum harmonic oscillator. Equation (8.51) has the solution

$$\phi_-(t, \mathbf{x}) = (\cos(Kt)\phi_-)(0, \mathbf{x}) + (\sin(Kt)\phi_+)(0, \mathbf{x})$$
$$+ \sqrt{2\hbar} \int_0^t \sin(K(t - s))\sqrt{g_{00}}g^{-\frac{1}{4}}dW_s.$$

8.7 Time-Dependent Gaussian State in Quantum Field Theory

We are mainly interested in this chapter in a functional representation of Gaussian states for free fields defined on a Riemannian manifold with a time-dependent metric. First, let us consider the simple case of the Gaussian state for a free field theory defined on the Minkowski space-time determined by the heuristic Hamiltonian (2.56) (its proper mathematical definition is achieved by the transformation (2.59) leading to the Hamiltonian (2.60))

$$\psi_t^\Gamma = A(t) \exp\left(-\frac{1}{2\hbar}\phi\Gamma_t\phi\right). \tag{8.52}$$

If we consider the model of the free quantum field in an infinite volume then in Eq. (8.52)

$$\phi\Gamma_t\phi = \int d\mathbf{x}d\mathbf{y}\phi(\mathbf{x})\phi(\mathbf{y})\Gamma_t(\mathbf{x}-\mathbf{y}) = \int d\mathbf{k}\phi(\mathbf{k})\phi(-\mathbf{k})\Gamma_t(\mathbf{k}),$$

where (we denote a function and its Fourier transform by the same letter)

$$\Gamma_t(\mathbf{x}) = (2\pi)^{-3}\int d\mathbf{k}\Gamma_t(\mathbf{k})\exp(i\mathbf{k}\mathbf{x})$$

and

$$\phi(\mathbf{x}) = (2\pi)^{-\frac{3}{2}}\int d\mathbf{k}\phi(\mathbf{k})\exp(i\mathbf{k}\mathbf{x}).$$

However, for problems discussed in this and subsequent sections a finite volume regularization is recommended. Then, we consider a cube of volume $\Omega = L^3$ and periodic fields with the period L

$$\Gamma_t(\mathbf{x}) = \Omega^{-1}\sum_\mathbf{n}\Gamma_t(\mathbf{n})\exp\left(i\frac{2\pi}{L}\mathbf{n}\mathbf{x}\right)$$

$$\phi(\mathbf{x}) = \Omega^{-\frac{1}{2}}\sum_\mathbf{n}\phi(\mathbf{n})\exp\left(i\frac{2\pi}{L}\mathbf{n}\mathbf{x}\right)$$

and

$$\phi\Gamma_t\phi = \int_\Omega d\mathbf{x}\int_\Omega d\mathbf{y}\phi(\mathbf{x})\phi(\mathbf{y})\Gamma_t(\mathbf{x}-\mathbf{y}) = \sum_\mathbf{n}\phi(\mathbf{n})\phi(-\mathbf{n})\Gamma_t(\mathbf{n}),$$

where the sum is over integers $\mathbf{n} \in Z^3$.

ψ_t is a solution of the Schrödinger equation

$$i\hbar\partial_t\psi_t = H_0\psi_t,$$ (8.53)

if (we consider solutions satisfying the condition $\Gamma(\mathbf{k}) = \Gamma(-\mathbf{k}) = \Gamma(k)$ where $k = |\mathbf{k}|$)

$$i\partial_t\Gamma - \Gamma^2 + \mathbf{k}^2 + m^2 = 0.$$ (8.54)

For A in Eq. (8.52) we have the formula

$$A(t) = \exp\left(-\frac{i}{2}\int d\mathbf{x}\int_{t_0}^t ds \int d\mathbf{k}\Gamma_s(\mathbf{k})\right).$$

$A(t)$ needs a volume cutoff. When we calculate correlation functions in the state ψ_t^Γ then the normalization factor $|A(t)|^2$ cancels. In a finite volume $A(t)$ is finite (then $\int d\mathbf{k} \to \sum_\mathbf{n}$ in the expression for $A(t)$). Note that if Γ is real then $|A| = 1$ and the factor $A(t)$ does not enter correlation functions. For the time independent solution of Eq. (8.53) $A(t)$ is just the phase corresponding to the ground state energy of the quantum field.

The Riccati equation (8.54) can be related to a linear second order differential equation. Let

$$u_t(\mathbf{k}) = \exp\left(i\int_{t_0}^t ds\Gamma(s,\mathbf{k})\right),$$ (8.55)

then

$$\frac{d^2u}{dt^2} + (\mathbf{k}^2 + m^2)u = 0.$$ (8.56)

We obtain Γ from

$$i\Gamma = u^{-1}\frac{du}{dt}.$$

The general solution of Eq. (8.56) can be expressed as

$$u = \alpha\cos(\omega t) + \delta\sin(\omega t),$$

where

$$\omega(\mathbf{k}) = \sqrt{m^2 + \mathbf{k}^2}.$$

Then

$$i\Gamma = \omega(-\alpha\sin(\omega t) + \delta\cos(\omega t))(\alpha\cos(\omega t) + \delta\sin(\omega t))^{-1}.$$

We can write the general solution also in a complex form

$$u = c_1\exp(i\omega t) + c_2\exp(-i\omega t),$$

then $c_2 = 0$ gives the ground state in Eq. (8.52). Another form of the solution of Eq. (8.56) is

$$u = \cos(\omega t + \alpha - i\nu),$$

where ν and α are arbitrary functions of $\mathbf{k}$. Then, Γ has the form similar to Eq. (8.14)

$$\Gamma_t = i\omega \tan(\omega t + \alpha - i\nu) = \\ \omega(\sinh(2\nu) + i\sin(2\omega t + 2\alpha))(\cosh(2\nu) + \cos(2\omega t + 2\alpha))^{-1}. \qquad (8.57)$$

The initial condition is written as $\Gamma_0 = i\omega \tan(\alpha - i\nu)$.

We consider the measures $d\sigma_t = d\phi|\psi_t^\Gamma|^2$ and $d\mu_0 = d\phi|\psi^g|^2$ where ψ^g is the ground state (2.58). These measures determine correlation functions of quantum fields. If the volume Ω is infinite then the measure $d\sigma_t$ is not continuous with respect to $d\mu_0$. This means that the states ψ_t^Γ cannot be realized in the Hilbert space $L^2(d\mu_0)$. However, for a finite volume Ω

$$d\sigma_t(\phi) = d\mu_0(\phi) \exp\left(-\tfrac{1}{2\hbar}\phi(\Gamma_t + \Gamma_t^* - 2\omega)\phi\right)\left(\det \tfrac{\Gamma_t + \Gamma_t^*}{2\omega}\right)^{\frac{1}{2}}, \qquad (8.58)$$

where

$$\det \frac{\Gamma_t + \Gamma_t^*}{2\omega} = \exp Tr \ln(1 - K)$$

with

$$K = (\cos(2\omega t + 2\alpha) - \exp(-2\nu))(\cosh(2\nu) + \cos(2\omega t + 2\alpha))^{-1} \qquad (8.59)$$

The variables ω, α and ν in general depend on $\mathbf{k}$. In the Fourier space they become multiplication operators. In a finite volume Ω the multiplication operator K depends on integers $\mathbf{n} \in Z^3$. We have for any operator $\mathcal{K}$ (depending on $\mathbf{n}$) being a multiplication operator in the discrete Fourier space

$$Tr\mathcal{K} = \int_\Omega d\mathbf{x}\mathcal{K}(\mathbf{x}, \mathbf{x}) = \Omega \sum_{\mathbf{n}} \mathcal{K}(\mathbf{n}).$$

It can be proved [222] that $|\det(1 - \mathcal{K})| \leq \exp|\mathcal{K}|_1$ where $|..|_1$ is the trace norm in the space of trace class operators. If we choose $\nu(\mathbf{n})$ such that

$$\cosh(2\nu(\mathbf{n})) > 1 + \epsilon_1 + \epsilon_2 \mathbf{n}^4,$$

where $\epsilon_1 > 0$ and $\epsilon_2 > 0$ then for a certain $R > 0$

$$|K|_1 < R\Omega \sum_{\mathbf{n}} (\mathbf{n}^2)^{-2} < \infty.$$

We could also expand $Tr \ln(1 - K)$ in powers of K. We can show that under our assumptions $Tr(K^n)$ is finite for $n \geq 1$ and the power series is convergent. The continuity of measures in our description of the functional representation of Sect. 2.5 (see the formulation and the results in [225], Chap. 3, Sect. 17 or [172], Chap. 2, Sect. 3) means that after the transformation of states $\psi_t^\Gamma \to (\psi^g)^{-1}\psi_t^\Gamma \equiv \hat{\psi}_t^\Gamma$ and operators $A \to (\psi^g)^{-1}A\psi^g$ we may work in the Hilbert space $L^2(d\mu_0)$ as $|\hat{\psi}_t^\Gamma|^2$ is integrable with respect to $d\mu_0$.

Let us note that the measure $d\sigma_t$ is also continuous with respect to $d\sigma_0$

$$d\sigma_t(\phi) = d\sigma_0(\phi) \exp\left(-\tfrac{1}{2\hbar}\phi(\Gamma_t + \Gamma_t^* - \Gamma_0 - \Gamma_0^*)\phi \right) \left(\det \tfrac{\Gamma_t + \Gamma_t^*}{\Gamma_0 + \Gamma_0^*} \right)^{\frac{1}{2}}, \quad (8.60)$$

where the determinant in Eq. (8.60) is of the form $\det(1 - \tilde{K})$ with

$$\tilde{K} = (\cos(2\alpha + 2\omega t) - \cos(2\alpha))(\cosh(2\nu) + \cos(2\alpha + 2\omega t))^{-1}.$$

It follows that we could choose $L^2(d\sigma_0)$ as the Hilbert space for QFT instead of $L^2(d\mu_0)$. This is relevant for QFT in a time-dependent metric as in this case there is no ground state to start with. In subsequent sections of this chapter we do not explicitly introduce the volume cutoff. It is understood that $d\sigma_t$ defining correlations on the phase space by means of the Wigner function and the Wigner-Weyl transform of Sect. 4.2 can define states in the sense of the algebraic QFT [132].

We restrict ourselves to a discussion of Gaussian Wigner functions (in quantum free field theories) which are of the form

$$p(\phi, \Pi) \equiv \exp\left(-\tfrac{1}{2}X^T G X \right)(\det G)^{\frac{1}{2}}, \quad (8.61)$$

where we define a column X so that $X^T = (\phi, \Pi)$. G is 2×2 matrix of operators with $\det G = \det(G_{11}G_{22} - G_{12}^2)$. G_{jk} are Hermitian operators defined by the kernels $G_{ij}(\mathbf{x} - \mathbf{y})$. The formal Wigner function (8.61) could be defined as the Gaussian measure $d\mu^W(\phi, \Pi)$ in the sense of Sect. 1.3 by the Fourier transform of $d\mu^W(\phi, \Pi) = d\phi d\Pi p(\phi, \Pi)$

$$Z[f, h] = \int d\mu^W(\phi, \Pi) \exp(i(X^T, Y)) = \exp\left(-\tfrac{1}{2}Y^T G^{-1} Y \right), \quad (8.62)$$

where $Y^T = (f, h), (X^T, Y) = (f, \phi) + (h, \Pi)$ and G^{-1} is the inverse matrix to G equal

$$G^{-1} = \left(G_{11}G_{22} - G_{12}^2 \right)^{-1} \begin{bmatrix} G_{22} & -G_{12} \\ -G_{12} & G_{11} \end{bmatrix}.$$

Note that from Eq. (8.62) there follows the normalization $\int d\mu^W(\phi, \Pi) = 1$. If $p(\phi, \Pi)$ is to be Gaussian then a pure state ψ must be of the form (8.52). We calculate the Wigner function of the density matrix $\rho(\phi, \phi') = \psi(\phi)^*\psi(\phi')$ for ψ of the form

(8.52). We decompose

$$\Gamma = A + iB$$

into the real and imaginary parts then (for $[A, B] = 0$)

$$G_{11} = \frac{2}{\hbar} A + 2\hbar B A^{-1} B$$

$$G_{22} = 2\hbar A^{-1}$$

$$G_{12} = G_{21} = -2\hbar A^{-1} B$$

In such a case Eq. (8.61) takes the form

$$p(\phi, \Pi) = \exp\left(-\tfrac{1}{2} X^T G X\right)$$
$$= \exp\left(-\tfrac{1}{\hbar}(\phi, A\phi) - \hbar(\Pi, A^{-1}\Pi) - \hbar(\phi, BA^{-1}B\phi) + 2\hbar(\Pi, A^{-1}B\phi)\right),$$

There is no determinant in $p(\phi, \Pi)$ because $\det G = \det(4) = const$ (we omit the constant normalization factor). In Sect. 9.10 we discuss the entropy of a Gaussian Wigner function. We show there that the entropy is equal to $-\frac{1}{2} \ln \det G$. Hence, the fact that $\det G = const$ means that pure Gaussian states have a constant entropy.

The states described in this section are the squeezed states of the scalar field [181, 231]. We could calculate correlation functions of the scalar field in the state (8.52) from the stochastic equation (8.41) using the formula (8.7). However, using the explicit solution of the Heisenberg equations of motion (see Eq. (4.26))

$$\phi_t(\phi, \Pi) = \cos(\omega t)\phi + \omega^{-1} \sin(\omega t)\Pi$$

we can compute the correlation functions (8.7) in the squeezed state directly in the Heisenberg picture. We can also use the Wigner function to compute

$$\tfrac{1}{2}(\psi^\Gamma, \Phi(\mathbf{x})\Phi_t(\mathbf{x}')\psi^\Gamma) + \tfrac{1}{2}(\psi^\Gamma, \Phi_t(\mathbf{x}')\Phi(\mathbf{x})\psi^\Gamma) = \int d\mu^W(\phi, \Pi)\phi(\mathbf{x})\phi_t(\phi, \Pi; \mathbf{x}'),$$

where on the rhs we have a classical correlation function. The result that the field correlation functions can be calculated as expectation values with respect to the Wigner measure $d\mu^W$ follows from the Wigner-Weyl correspondence [89]. For fields with an interaction the solution of the Heisenberg equations of motion is unknown. The correlation functions can be calculated using the Feynman formula (8.23).

At the end of this section let us note that the Schrödinger equation (8.53) for the free field (equivalent to Eq. (8.54)) has a solution for $\omega(k)^2 = -(\mathbf{k}^2 + m^2) = -\nu^2$ (this is an analog of the upside-down oscillator of Sect. 8.2; in fact, there is a solution for $-\gamma(\mathbf{k}^2 + m^2)$ with any complex γ). Such an equation would describe a model with the Lagrangian (after an addition of the potential V)

$$\mathcal{L} = \frac{1}{2}((\partial_0\phi)^2 + (\nabla\phi)^2 + m^2\phi^2) - V(\phi)$$

which is Euclidean invariant. Then, the energy density

$$H(x) = \frac{1}{2}((\partial_0\phi)^2 - (\nabla\phi)^2 - m^2\phi^2) + V(\phi)$$

is Lorentz invariant. Such a model could be used for a description in a real time of the theories with negative potentials discussed in Sect. 6.1.

In these models the pressure is [59, 195] $P = \frac{1}{2}\Pi^2 + \frac{1}{2}((\nabla\phi)^2 + m^2\phi^2) - V(\phi)$. Hence, the ratio $w = PH^{-1} < -1$ for $V \le 0$ and $\Pi \simeq 0$. Models with $w < -1$ are interesting in cosmology [195] as they lead to cosmological evolutions with an increasing dark energy (in an accelerated expansion).

The solution of the Schrödinger equation (8.53) is of the form (8.52) with

$$i\Gamma = \nu(\alpha\sinh(\nu t) + \delta\cosh(\nu t))(\alpha\cosh(\nu t) + \delta\sinh(\nu t))^{-1},$$

where $\nu(\mathbf{k}) = \sqrt{m^2 + \mathbf{k}^2}$. A particular solution with $\alpha = \delta$ is

$$\psi_t^\Gamma = Z(t)^{-1}\exp\left(\frac{i}{2\hbar}\phi\nu\phi\right),$$

where

$$Z(t) = \exp\left(\frac{1}{2}t\int dxd\mathbf{k}\nu\right).$$

Then, the solution of the Schrödinger equation (8.53) with the initial condition

$$\psi = \exp\left(\frac{i}{2\hbar}\phi\nu\phi\right)\chi$$

is

$$\psi_t(\phi) = Z(t)^{-1}\exp\left(\frac{i}{2\hbar}\phi\nu\phi\right)E[\chi(\phi_t(\phi))],$$

where $\phi_t(\phi)$ is the solution of the stochastic equation

$$d\phi = -\nu\phi dt + \sigma dW_s$$

with $\nu = \sqrt{-\Delta + m^2}$, i.e.,

$$\phi_t(\phi) = \exp(-\nu t)\phi + \sigma\int_0^t \exp(-\nu(t-s))dW_s.$$

$Z(t)$ is infinite (without cutoffs) but it cancels in calculations of a ratio of expectation values.

We can derive a solution of the Schrödinger equation for the Hamiltonian $H = H_0 + V$ where V is the polynomial potential discussed at the end of Sect. 6.1 (Eq. (6.3)) with the scaled ϕ (in the sense of Eq. (5.39))

$$\psi_t(a\sigma\phi) = Z(t)^{-1} \exp(-\tfrac{a^2}{2}\phi\nu\phi)$$
$$E\left[\exp\left(-\tfrac{i}{\hbar}\int_0^t V(\phi_s(a\sigma\phi))d\mathbf{x}ds\right)\chi(\phi_t(a\sigma\phi))\right]. \tag{8.63}$$

We consider as an example the polynomial interactions in two dimensions of the form

$$V(\phi) = -\mu^2 : \phi^2 : +\lambda : \phi^4 : +g : \phi^6 : .$$

We introduce the volume cutoff ($x \in [0, L]$), the ultraviolet cutoff ϵ of Sect. 6.3, $\kappa_\epsilon(k) = 1 + \epsilon k^4$, the regularization $\nu_\epsilon = \kappa_\epsilon(k)\nu$ and regularized fields ϕ^ϵ as solutions of the stochastic equation with the regularized ν_ϵ. Then, we have

$$\left|\exp\left(-\frac{i}{\hbar}\int_0^t ds \int_0^L dx : V(\phi_s^\epsilon(a\sigma\phi)) :\right)\right| \leq \exp\left(tLK\left(\ln\frac{1}{\epsilon}\right)^3\right)$$

with a certain constant K. This estimate follows from an estimate of the lower bound of the sixth order polynomial whose coefficients at the $2n$th order term behave as $< \phi_\epsilon^2(x) >^{3-n}$ where $n = 1, 2$. We can prove as in [125] (Sect. VII.4) by means of the Duhamel expansion that when we remove the regularization then the limit

$$\lim_{\epsilon\to 0} E\left[\exp\left(-\frac{i}{\hbar}\int_0^t ds \int_0^L dx : V(\phi_s^\epsilon(a\sigma\phi)) :\right)\right]$$

exists. In the same way we could show that the limit of the Feynman integral

$$\lim_{\epsilon\to 0} E\left[\exp\left(-\frac{i}{\hbar}\int_0^t ds \int_0^L dx : V(\phi_s^\epsilon(a\sigma\phi)) :\right)\chi(a\sigma\phi_t)\right]$$

exists for a certain set of initial wave functions $\chi(\phi)$. The method of the Duhamel expansion is based on the estimate (for natural $p \geq 1$)

$$E\left[\left|: V(\phi_s^\epsilon(a\sigma\phi)) : - : V(\phi_s(a\sigma\phi)) :\right|^p\right] < C^p \epsilon^{\beta p} p^{3p}$$

with certain positive constants C and β. So that the exponentially growing estimate $\exp\left(tLK(\ln\frac{1}{\epsilon})^3\right)$ is compensated by the multiplication by $C^p\epsilon^{\beta p}p^{3p}$ in the Duhamel expansion. As a result the limit $\epsilon \to 0$ exists. As the initial wave function we may consider the plane waves $\exp(ik(\phi, f))$ or some localized wave functions as ,e.g.,$\exp(ik(\phi, f))\exp(-(\phi, f)^4)$.

In two-dimensional space-time the proof of ultraviolet finiteness of the polynomial models of Sect. 6.1 (with the normal ordering) will proceed in the same way as in $P(\phi)_2$ models [124, 125] because the estimates for the functional integration in Eq. (8.63) used in the proofs are the same as in the models of [124, 125] after an application of the complex scaling $\phi \to a\sigma\phi$. The model with an inverted sign of the ω^2 and the wave functions $\psi_t(a\sigma\phi)$ could be treated as the complex scaling technique in QFT analogous to the one in quantum mechanics [205, 223]. The benefit of such a scaling is that the evolution can be discussed in the real time instead of the imaginary time of Euclidean field theory. After the complex scaling we can also approach in the rigorous functional integration formalism the potentials which are unbounded from below (see [146] for a study of the formula (8.63) in QFT in two and more dimensions).

8.8 Free Field in an Expanding Universe

We discuss the functional integral approach to quantum field theory on a manifold in a complete analogy to the Schrödinger picture in quantum mechanics (see a similar approach in [88, 130, 161, 178]). In contradistinction to the Heisenberg picture we insist on states and their time evolution. In quantum field theory with a static metric we have the Fock ground state as the unique ground state. With a general time-dependent metric we do not expect that a time-independent ground state exists.We concentrate on the time evolution of Gaussian states and expectation values of observables in these states. The Gaussian states are distinguished in quantum theory as they define, by means of the Wigner function, a probability measure on the phase space. We can use the Wigner distribution to describe the expectation values of observables as explained in Sects. 4.2 and 8.7.

We restrict ourselves in this section to a spatially flat (pseudo) Riemannian manifold $\mathcal{M}$ with the metric (this may be defined only on a submanifold $\mathcal{M}_0 \subset \mathcal{M}$)

$$ds^2 = g_{\mu\nu}dx^\mu dx^\nu = dt^2 - a(t)^2 d\mathbf{x}^2.$$

The free field satisfies the Klein-Gordon equation (4.19)

$$\partial_t^2\Phi - a^{-2}\Delta\Phi + 3\mathcal{H}\partial_t\Phi + m^2\Phi = 0, \tag{8.64}$$

where $\mathcal{H} = a^{-1}\partial_t a$. If Φ_1 and Φ_2 are different solutions of Eq. (8.64) then from Eq. (8.64) it follows

$$\partial_t \int dx a^3 \left(\Phi_1 \partial_t \Phi_2 - \Phi_2 \partial_t \Phi_1 \right) = 0$$

If we write

$$\Phi(t, \mathbf{x}) = a^{-\frac{3}{2}} \phi(t, \mathbf{x}) \tag{8.65}$$

then Eq. (8.64) can be rewritten as

$$\partial_t^2 \phi - \omega_a^2 \phi = 0, \tag{8.66}$$

where

$$\omega_a^2 = \frac{9}{4}\mathcal{H}^2 + \frac{3}{2}\partial_t \mathcal{H} + a^{-2}\Delta - m^2. \tag{8.67}$$

The Lagrangian for the system (8.64) is

$$\mathcal{L} = \frac{1}{2}\sqrt{g}\left(g^{\mu\nu} \partial_\mu \Phi \partial_\nu \Phi - m^2 \Phi^2 \right).$$

The canonical momentum

$$\Pi = \sqrt{g}\partial_t \Phi. \tag{8.68}$$

The Hamiltonian

$$H = \frac{1}{2} \int dx \left(a^{-3} \Pi^2 + a(\nabla \Phi)^2 + m^2 a^3 \Phi^2 \right) \tag{8.69}$$

or using the canonical representation (2.55)

$$H = \frac{1}{2} \int dx \left(-\hbar^2 a^{-3} \frac{\delta^2}{\delta \Phi(\mathbf{x})^2} + a(\nabla \Phi)^2 + m^2 a^3 \Phi^2 \right). \tag{8.70}$$

Let ψ_t^Γ be a Gaussian solution (8.52) of the Schrödinger equation (8.53) with the Hamiltonian H of Eq. (8.70). Then, according to the formalism of Sects. 8.1 and 8.7 a general solution ψ_t can be expressed in the form $\psi_t = \psi_t^\Gamma \chi_t$ where χ_t satisfies the equation

$$\partial_t \chi = \int dx \left(\frac{i\hbar}{2} a^{-3} \frac{\delta^2}{\delta \Phi(\mathbf{x})^2} - i a^{-3} \Gamma \Phi \frac{\delta}{\delta \Phi(\mathbf{x})} \right) \chi \tag{8.71}$$

(χ is assumed to belong to the Hilbert space $L^2(d\sigma_t)$ with the well-defined Gaussian measure $d\sigma_t(\Phi) = |\psi_t^\Gamma|^2 d\Phi$). The stochastic equation (8.5) reads (with the modification resulting from $a \neq 1$)

$$d\Phi_s = -i a^{-3}(t - s)\Gamma(t - s)\Phi_s ds + a^{-\frac{3}{2}}(t - s)\sqrt{i\hbar} dW_s. \tag{8.72}$$

The solution of Eq. (8.72) with the initial condition Φ at $s = t_0$ is

$$\Phi_s = \exp(-i \int_{t_0}^s (a^{-3}\Gamma)(t - \tau)d\tau)\Phi$$
$$+\sqrt{i\hbar} \int_{t_0}^s \exp(-i \int_{\tau}^s (a^{-3}\Gamma)(t - \tau')d\tau')a^{-\frac{3}{2}}(t - \tau)dW_\tau. \tag{8.73}$$

There remains to determine $\Gamma(t)$. In Eq. (8.52) Γ is an operator with an integral kernel $\Gamma(\mathbf{x}, \mathbf{y})$. We can derive an equation for this operator demanding that (8.52) is the solution of the Schrödinger equation (8.53) with the Hamiltonian (8.70). Then, we obtain a generalization of Eq. (8.54)

$$i\partial_t \Gamma - a^{-3}\Gamma^2 + (-a\Delta + m^2a^3) = 0. \tag{8.74}$$

We can relate this non-linear Riccati equation to a linear second order differential equation if we introduce the operator u

$$u = \exp\left(i \int^t ds a(s)^{-3}\Gamma(s)\right). \tag{8.75}$$

Then

$$\frac{d^2u}{dt^2} + 3\mathcal{H}\frac{du}{dt} + (-a^{-2}\Delta + m^2)u = 0. \tag{8.76}$$

This is the same equation as the Klein-Gordon equation (8.64). In the case of the homogeneous metric the integral kernel of Γ can be expressed by its Fourier transform

$$\Gamma(\mathbf{x} - \mathbf{y}) = (2\pi)^{-3} \int d\mathbf{k}\Gamma(\mathbf{k}) \exp(i\mathbf{k}(\mathbf{x} - \mathbf{y})).$$

We consider solutions satisfying the condition $\Gamma(\mathbf{k}) = \Gamma(-\mathbf{k}) = \Gamma(k)$ where $k = |\mathbf{k}|$. Then, in Fourier transforms Eq. (8.74) reads

$$i\partial_t \Gamma - a^{-3}\Gamma^2 + a\mathbf{k}^2 + m^2a^3 = 0. \tag{8.77}$$

For A in Eq. (8.52) we have the formula

$$A(t) = \exp\left(-\frac{i}{2} \int d\mathbf{x} \int_{t_0}^t ds \int d\mathbf{k} a^{-3}\Gamma(s, \mathbf{k})\right). \tag{8.78}$$

Equation (8.76) in momentum space reads

$$\frac{d^2u}{dt^2} + 3\mathcal{H}\frac{du}{dt} + (a^{-2}\mathbf{k}^2 + m^2)u = 0. \tag{8.79}$$

We obtain Γ from

$$i\Gamma = u^{-1}\frac{du}{dt}a^3. \tag{8.80}$$

As in Sect. 8.7 we can make $A(t)$ in Eq. (8.78) finite if we restrict the model to a box of finite volume $\Omega = L^3$ and impose periodic boundary conditions on the Laplacian in Eq. (8.76). In such a case $\mathbf{k}$ in Eqs. (8.77) and (8.79) takes discrete values $\frac{2\pi}{L}\mathbf{n}$ (then there is a sum over $\mathbf{n}$ instead of the integral in Eq. (8.78)).

In terms of

$$v = a^{\frac{3}{2}}u$$

we obtain a Schrödinger-like equation (t plays the role of a coordinate, $\omega_a^2(t)$ is the potential)

$$\frac{d^2v}{dt^2} - \omega_a^2 v = 0, \tag{8.81}$$

where ω_a^2 is defined in Eq. (8.67).

Let us note that from Eq. (8.79) it follows

$$\partial_t(a^3(u\partial_t u^* - u^*\partial_t u)) = 0.$$

We choose the normalization

$$u(\mathbf{k})\partial_t u^*(\mathbf{k}) - u^*(\mathbf{k})\partial_t u(\mathbf{k}) = ia^{-3}.$$

A real field Φ solving the Klein-Gordon equation (8.64) can be expanded in the Fourier series

$$\begin{aligned}
\Phi_t(\mathbf{x}) &\equiv (2\pi)^{-\frac{3}{2}} \int d\mathbf{k}\, \exp(i\mathbf{kx})\Phi_t(\mathbf{k}) \\
&= (2\pi)^{-\frac{3}{2}} \int d\mathbf{k}\, \exp(i\mathbf{kx})(b(\mathbf{k})u_t(\mathbf{k}) + b^*(-\mathbf{k})u_t^*(-\mathbf{k})).
\end{aligned} \tag{8.82}$$

We quantize the field (8.82) in the Fock space choosing b as an annihilation operator and b^* as the creation operator, Then, in terms of fields

$$b(\mathbf{k}) = -\frac{i}{\sqrt{2}}a^3\Phi(\mathbf{k})\partial_t u(\mathbf{k}) + \sqrt{2}u(\mathbf{k})\hbar\frac{\delta}{\delta\Phi^*(\mathbf{k})}, \tag{8.83}$$

and

$$b^+(\mathbf{k}) = \frac{i}{\sqrt{2}}a^3\Phi^*(\mathbf{k})\partial_t u^*(\mathbf{k}) + \sqrt{2}u^*(\mathbf{k})\hbar\frac{\delta}{\delta\Phi(\mathbf{k})}.$$

By direct calculations with ψ_t^Γ of Eq. (8.52) and Γ_t defined in Eq. (8.80) we have

$$b(\mathbf{k})\psi_t^\Gamma = 0. \tag{8.84}$$

In the Hilbert space $L^2(|\psi_t^\Gamma|^2 d\Phi)$ as discussed in Sect. 2.5

$$\tilde{b} = (\psi_t^\Gamma)^{-1}b\psi_t^\Gamma = \sqrt{2}u(\mathbf{k})\hbar\frac{\delta}{\delta\Phi^*(\mathbf{k})}$$

so that $\tilde{b}(\mathbf{k})1 = 0$. Using the normalization of $u(\mathbf{k})$ we check that

$$[b(\mathbf{k}), b^+(\mathbf{k}')] = [\tilde{b}(\mathbf{k}), \tilde{b}^+(\mathbf{k}')] = \hbar\delta(\mathbf{k} - \mathbf{k}').$$

The formula (8.82) expressing the field in terms of creation and annihilation operators can also be written in the form which is a realization of Eq. (4.26)

$$\Phi_t(\Phi, \Pi) = \mathcal{K}_t\Phi + \mathcal{L}_t\Pi \qquad (8.85)$$

as an expression for the solution of the Cauchy problem [54] for the hyperbolic equation in terms of the kernels $\mathcal{K}_t$ and $\mathcal{L}_t$. From such a formula (the kernels can be expressed by the solution u of Eq. (8.79)) we can derive the field correlation functions of Φ_t because the expectation values of Φ and Π are expressed as expectation values with respect to the Wigner measure $d\mu^W(\Phi, \Pi)$ on the phase space as explained at the end of Sect. 8.7. The expression of the evolution of the wave function with an interaction potential V can be written in terms of the random field (8.73) (as in the Feynman-Kac formula Eq. (8.35)).

The Hilbert space $L^2(|\psi_t^\Gamma|^2 d\Phi)$ is well-defined if $\Gamma + \Gamma^* > 0$. Γ is defined by Eq. (8.80) and a priori the fulfillment of the condition $\Gamma + \Gamma^* > 0$ is not guaranteed. In refs. [22, 24, 57, 239] the construction of quantum fields on a manifold is formulated in geometric terms. The existence of the Gaussian measure (defined by a positive bilinear form) is imposed as a condition on the symplectic form and the complex structure. In the next section in the case of de Sitter space we show that we can choose the solution u of Eq. (8.79) such that Γ of Eq. (8.80) satisfies the condition $\Gamma + \Gamma^* > 0$. For a similar approach to the Schrödinger picture for fields on a manifold with a time-dependent metric see [88, 130, 178].

8.9 Free Field in De Sitter Space

De Sitter space can be described as the hyperboloid in 5 dimensional Minkowski space

$$z_0^2 - z_1^2 - z_2^2 - z_3^2 - z_4^2 = -R^2.$$

We can choose coordinates on the half of de Sitter space such that the induced metric is spatially flat [42, 46] with $a(t) = \exp(\mathcal{H}t)$ where $\mathcal{H} = \frac{1}{R}$ is constant. The Schrödinger equation for a Hamiltonian quadratic in canonical variables has a Gaussian solution (8.52). If $a = \exp(\mathcal{H}t)$ then the solution of Eq. (8.81) is the cylinder function Z_ν [29, 115]

$$v = Z_\nu\left(\frac{k}{\mathcal{H}}\exp(-\mathcal{H}t)\right), \qquad (8.86)$$

where

$$\nu = \frac{3}{2}\sqrt{1 - \frac{4m^2}{\mathcal{H}^2}}.$$ (8.87)

The general solution is a linear combination of cylinder functions

$$\beta(k)Z_\nu\left(\frac{k}{\mathcal{H}}\exp(-\mathcal{H}t)\right) + \gamma(k)Z_\nu^*\left(\frac{k}{\mathcal{H}}\exp(-\mathcal{H}t)\right)$$

However, as we discuss later, if the solution is to be asymptotically an incoming wave then we must choose as the solution either Z_ν or Z_ν^*.

The solution of Eq. (8.77) in terms of v (8.86) is

$$i\Gamma = a^3\frac{d}{dt}\ln\left(a^{-\frac{3}{2}}Z_\nu\left(\frac{k}{\mathcal{H}}\exp(-\mathcal{H}t)\right)\right).$$ (8.88)

Then, the stochastic equation reads

$$d\Phi_s = -\partial_t\ln\left(a^{-\frac{3}{2}}(t-s)Z_\nu\left(\tfrac{k}{\mathcal{H}}\exp(\mathcal{H}s-\mathcal{H}t)\right)\right)\Phi_s ds + a^{-\frac{3}{2}}(t-s)\sqrt{i\hbar}dW_s$$ (8.89)

which has the solution (with the initial condition Φ at t_0)

$$\Phi_s = a^{-\frac{3}{2}}(t-s)Z_\nu\left(\tfrac{k}{\mathcal{H}}\exp(-\mathcal{H}t+\mathcal{H}s)\right)a^{\frac{3}{2}}(t-t_0)$$
$$\times\left(Z_\nu(\tfrac{k}{\mathcal{H}}\exp(-\mathcal{H}t+\mathcal{H}t_0))\right)^{-1}\Phi$$
$$+\sqrt{i\hbar}a^{-\frac{3}{2}}(t-s)\left(Z_\nu(\tfrac{k}{\mathcal{H}}\exp(\mathcal{H}t-\mathcal{H}s))\right)\int_{t_0}^s Z_\nu\left(\tfrac{k}{\mathcal{H}}\exp(-\mathcal{H}t+\mathcal{H}\tau))\right)^{-1}dW_\tau.$$ (8.90)

We consider the Schrödinger equation (8.53) on the interval $[s, t]$. It defines the unitary evolution $\Psi_t = U_{t,s}\Psi_s$. We can define Heisenberg evolution of fields

$$\hat{\Phi}_t = U_{t,s}^+\Phi_s U_{t,s}.$$ (8.91)

The solution of the stochastic equation determines the field correlation functions (where ψ^Γ is defined by Γ in Eq.(8.52))

$$(\psi_0^\Gamma, F_1(\hat{\Phi}_t)F_2(\hat{\Phi})\psi_0^\Gamma) = \int d\Phi|\psi_t^\Gamma(\Phi)|^2 F_1(\Phi)E\left[F_2\left(\Phi_t(\Phi)\right)\right].$$ (8.92)

From Eqs. (8.91)–(8.92) (under the assumption that the initial value is at $s = 0$)

$$(\psi_0^\Gamma, \hat{\Phi}_t\hat{\Phi}\psi_0^\Gamma) = \int d\Phi|\psi_t^\Gamma(\Phi)|^2\Phi E\left[\Phi_t(\Phi)\right]$$
$$= (\Gamma(t) + \Gamma(t)^*)^{-1}Z_\nu(\tfrac{k}{H})a^{\frac{3}{2}}(t)\left(Z_\nu(\tfrac{k}{H}\exp(-Ht))\right)^{-1}\delta(k+k')$$ (8.93)

Let us calculate (with $z = \frac{k}{\mathcal{H}}\exp(-\mathcal{H}t)$)

$$i(\Gamma(t) + \Gamma(t)^*) = a^3 Hz(Z_\nu^* Z_\nu)^{-1}\left(\tfrac{d}{dz}Z_\nu^* Z_\nu - \tfrac{d}{dz}Z_\nu Z_\nu^*\right)$$
$$= W(Z, Z^*)a^3 \mathcal{H}z(Z_\nu^* Z_\nu)^{-1},$$

where the Wronskian

$$W(Z, Z^*) = Z\partial_z Z^* - Z^*\partial_z Z.$$

The general solution Z_ν of Eq. (8.81) is a sum of Hankel functions [29, 115]

$$Z_\nu = \beta(k)H_\nu^{(1)\prime} + \gamma(k)H_\nu^{(2)}. \tag{8.94}$$

The Hankel functions have the asymptotic behavior in the form of plain waves

$$H_\nu^{(2)}(k\eta) \simeq (2k)^{-\frac{1}{2}}\exp(-ik\eta)$$

for a large $|\eta|$, $H^{(1)}(k\eta) \simeq (2k)^{-\frac{1}{2}}\exp(ik\eta))$. In the argument of Z_ν in Eq. (8.86) $-\frac{1}{\mathcal{H}}\exp(-\mathcal{H}t) = \int dt a(t)^{-1} = \eta$, where η is the conformal time. Hence, the Hankel functions behave for the remote past (or when $\mathcal{H} \to 0$) as plain waves. For the same reason the quantum fields (8.82) behave for $t \to -\infty$ as free fields in the Minkowski space.

$\Gamma + \Gamma^*$ in the exponent of $|\psi^\Gamma|^2$ may have any sign if Z_ν is a general linear combination (8.94) of Hankel functions. We have for $\beta = 0$ and $\gamma = 1$

$$W(H^{(2)}, H^{(2)*}) = i\frac{4}{\pi}z^{-1}. \tag{8.95}$$

Then, $\Gamma(t) + \Gamma(t)^*$ is positive so that $|\psi_t^\Gamma|^2$ in Eq. (8.93) integrable confirming the formula (8.99) with

$$\int d\Phi |\psi_t^\Gamma(\Phi)|^2 \Phi(\mathbf{k})\Phi(\mathbf{k}') = \hbar(\Gamma + \Gamma^*)^{-1}\delta(\mathbf{k} + \mathbf{k}'). \tag{8.96}$$

Using the expression for $\Gamma + \Gamma^*$ from the formula below Eq. (8.93) we obtain

$$(\psi_0^\Gamma, \hat{\Phi}_t^\varrho \hat{\Phi}^\varrho \psi_0^\Gamma) = Z_\nu(\tfrac{k}{H})a^{-\frac{3}{2}}(t)Z_\nu^*(\tfrac{k}{H}\exp(-Ht))\delta(k + k'). \tag{8.97}$$

The result (8.97) for correlation functions coincides with the formula for quantum fields in de Sitter space quantized in the Fock space [46]. For general theory of quantum fields on the Sitter space see [42].

We can write the first order stochastic equation (8.72) in the form of a random wave equation for the field ϕ of Eq. (8.65). By means of a direct differentiation of Eq. (8.72) and using Eq. (8.74) we obtain

$$\partial_s^2 \phi(s, \mathbf{x}) - \omega_a^2(s)\phi(s, \mathbf{x})$$
$$= \sqrt{i\hbar}\partial_s^2 W - \left(-\tfrac{3}{2}\mathcal{H} + ia^{-3}(s)\Gamma(s)\right)\sqrt{i\hbar}\partial_s W. \tag{8.98}$$

We could decompose Eq. (8.98) into the real and imaginary components as in Eq. (8.51) but the noise on the rhs is more complicated here because Γ is a complex function. It is interesting that quantum field on a manifold as in Eqs. (8.51) and (8.98) can be represented as a solution of a random wave equation. From Sect. 8.1 it can be seen that the statement remains true for the scalar fields with an interaction. However, in these more demanding models the noise cannot be so easily determined. This is already seen when comparing Eqs. (6.32), (8.51) and (8.98). In the latter case in order to determine the noise we must find Γ from Eq. (8.74).

The benefit of the stochastic representation of fields on a manifold consists in the possibility of an application of the Feynman-Kac formula for a time evolution with an interaction $V(\Phi)$. Then, the solution of the Schrödinger equation is $\psi_t = \psi_t^\Gamma \chi_t$ where

$$\chi_t = E\left[\exp\left(-\frac{i}{\hbar} \int_0^t V(\Phi_s)dxds \right) \chi(\Phi_t) \right].$$

There is a Feynman-Kac formula for quantum fields in the Fock space in the framework of non-commutative quantum probability [5] (see also [6, 85]). However, the use of random fields may be beneficial for numerical computation of expectation values in comparison with the operator approach.

8.10 Interference of Classical and Quantum Waves

At the end of this chapter we discuss the interference phenomena. The interference cannot be treated in the Euclidean formulation. In the classical Young interference experiment we consider a superposition of two plain waves

$$\psi = \exp(i\mathbf{k}_1\mathbf{x}) + \exp(i\mathbf{k}_2\mathbf{x}). \tag{8.99}$$

Then, the intensity of the light on the screen

$$P_{12} = |\exp(i\mathbf{k}_1\mathbf{x}) + \exp(i\mathbf{k}_2\mathbf{x})|^2 \tag{8.100}$$

shows maxima and minima depending on $\mathbf{k}_1\mathbf{x}$ and $\mathbf{k}_2\mathbf{x}$. There is a version of the Young experiment when the beam is split into two beams which are going different ways and then they meet and interfere on the screen. The classical plane wave evolves as

$$\psi^{(1)}(\mathbf{x}) = \int \mathcal{G}(t_1, \mathbf{x}; \mathbf{y}) \exp(i\mathbf{k}\mathbf{y})d\mathbf{y}, \tag{8.101}$$

where $\mathcal{G}(t, \mathbf{x}; \mathbf{y})$ is the Green function for the classical wave equation. The other wave $\psi^{(2)}(\mathbf{x})$ makes a longer way in time t_2. The interference is determined by $|\psi^{(1)}(\mathbf{x}) + \psi^{(2)}(\mathbf{x})|^2$.

In quantum mechanics the Green function $\mathcal{G}(t, \mathbf{x}; \mathbf{y}) = (\exp(-\frac{i}{\hbar}Ht))(\mathbf{x}, \mathbf{y})$ has been studied in Sect. 5.1. We could consider a more realistic problem of a propagation and interference of Gaussian beams in quantum mechanics [220]. After an evolution in time t the Gaussian analog of Eq. (8.99) is

$$\psi_t(\mathbf{x})$$
$$= \int \mathcal{G}(t, \mathbf{x}; \mathbf{y})\Big(\exp(i\mathbf{k}_1\mathbf{y})\exp(-\tfrac{1}{2\hbar}\Gamma_0\mathbf{y}^2) + \exp(i\mathbf{k}_2\mathbf{y})\exp(-\tfrac{1}{2\hbar}\Gamma_0\mathbf{y}^2)\Big)d\mathbf{y}. \tag{8.102}$$

According to Sect. 8.2 for a quadratic potential

$$\psi_t(\mathbf{x}) = A(t)\exp(-\frac{1}{2\hbar}\Gamma_t\mathbf{x}^2)E\Big[\exp(i\mathbf{k}_1\mathbf{q}_t(\mathbf{x})) + \exp(i\mathbf{k}_2\mathbf{q}_t(\mathbf{x}))\Big]. \tag{8.103}$$

The probability density on the screen is equal to $|\psi_t(\mathbf{x})|^2$. Instead of the interference of of the plain waves $\exp(i\mathbf{k}\mathbf{x})$ (8.99) we shall have the interference of the stochastic plain waves (8.103). The stochastic path is a solution of the equation

$$d\mathbf{q} = -i\Gamma_{t-s}\mathbf{q}_s ds + \sqrt{i\hbar}dw_s. \tag{8.104}$$

The solution is

$$\mathbf{q}_s(\mathbf{x}) = \exp(-i\int_{t_0}^{s}\Gamma(t-\tau)d\tau)\mathbf{x} + \sqrt{i\hbar}\int_{t_0}^{s}\exp(-i\int_{\tau}^{s}\Gamma(t-\tau')d\tau')dw_\tau. \tag{8.105}$$

For an oscillator in a coherent state $\Gamma = \omega$ we have

$$\mathbf{q}_t(\mathbf{x}) = \exp(-i\omega(t-t_0))\mathbf{x} + \sqrt{i\hbar}\int_{t_0}^{t}\exp(-i\omega(t-\tau))dw_\tau. \tag{8.106}$$

The interference of the waves (8.99) depends on $\exp(i(\mathbf{k}_1 - \mathbf{k}_2)\mathbf{x})$ whereas the interference of quantum beams (as follows from Eq. (8.103)) is time dependent and depends on $\exp(i\mathbf{k}_1\mathbf{q}_t(\mathbf{x}) - i\mathbf{k}_2\mathbf{q}_t^*(\mathbf{x}))$ where $\mathbf{q}_t^*$ is a complex conjugation of an independent version of $\mathbf{q}_t$.

Interference of light in QFT according to Glauber [110] depends on correlation functions in Gaussian states of the quantum electric field. These correlations can be expressed by correlations of random fields (cp. Eq. (8.92)) as will be discussed in the next chapter.

8.11   Exercises

8.1 Calculate $E[q_t q_s]$ in Eq. (8.26).
8.2 Show that from Eq. (8.49) it follows the stochastic wave equation (8.51).

8.3 Check Eq. (8.19) solving is Eq. (8.12) for the initial condition $\Gamma_0 = \mp i \nu x^2 + \gamma x^2$
(see Sect. 8.7) and comparing the solution with the Gaussian integral (8.19). In
order to calculate the expectation value of a Gaussian χ use the representation

$$\chi(x) = \exp(-\frac{a}{2}x^2) = (2\pi a)^{-\frac{1}{2}} \int dp \exp\left(-\frac{p^2}{2a}\right) \exp(ipx).$$

$E[\exp(iq_t(x))]$ can be calculated similarly as in Eq. (6.34).

8.4 Show the bound

$$|\exp(: V(\phi_\epsilon) :)| \le \exp\left(K\left(\ln\frac{1}{\epsilon}\right)^3\right)$$

in Eq. (8.63) as discussed at the end of Sect. 8.7. Hint: Estimate the lower
bound for an even sixth order polynomial $P(x)$ with a coefficient 1 at x^6 and the
coefficients $|\ln \epsilon|^{3-r}$ at x^{2r} (resulting from the normal ordering). Use this bound
to prove the convergence of the Duhamel perturbation series (see [125]).

8.5 Prove Eq. (8.77) for Γ.

8.6 Derive the formulas (8.58) and (8.60) using a finite dimensional approximation,
e.g., in the form of a Fourier expansion of fields in a finite volume with periodic
boundary conditions.

8.7 Prove that Eq. (8.73) is the solution of Eq. (8.72) and calculate
$\int d\Phi |\psi_t^\Gamma(\Phi)|^2 E[\Phi_t(\Phi)\Phi]$.

8.8 Show that b in Eq. (8.83) is an annihilation operator

8.9 Discuss the field expansion (8.82) and show that from the canonical commutation
relation $[\Phi, \Pi]$ with Π of Eq. (8.68) there follows that b^+ and b satisfy the
commutation relations for creation-annihilation operators.

8.10 Using Eq. (8.95) calculate the correlation function (8.97).

Chapter 9
An Interaction with a Quantum Electromagnetic Field

Abstract A quantization of an interaction with the electromagnetic field poses some difficulties. The problem is that in order to describe such interactions the field strength variables are not sufficient. We need the vector potentials. The problem arises with the covariant quantization of the four-potentials. We must choose the correct number of degrees of freedom for quantization. We can distinguish physical components imposing a gauge condition. In this chapter we work in various gauges and postpone the proof till Chap. 11 of the fact that the physics does not depend on the choice of gauge. We first quantize the electromagnetic field defining the functional integral by an introduction of a particular gauge fixing (the ξ-gauge). The functional integral is Lorentz (or Euclidean) invariant. However, the corresponding Wightman functions (or Euclidean Schwinger functions) do not satisfy the positivity requirement. These positivity property is satisfied in the $A_0 = 0$ gauge. We define the Higgs model and represent the correlation functions of this model as a Gibbs ensemble of polymers. In the subsequent sections we discuss a particle in a quantum electromagnetic field quantizing the field in the radiation gauge. We calculate the transition amplitudes and the density matrix in an environment of photons. Then, we study the particle motion in quantum electromagnetic field working in the Heisenberg picture. In such a description the creation and annihilation operators of the electromagnetic field play the role of the quantum noise (quantum Brownian motion). We derive a stochastic equation describing the radiation damping of the electron motion in the photon thermal state. The photon noise leads to the decoherence. In the relation to noise we discuss the notion of entropy in quantum field theory. The functional integral approach leads to a definition of entropy similar to the one in classical random systems. We calculate the entropy of a Gaussian Wigner function.

A quantization of an interaction with the electromagnetic field poses some difficulties. The problem is that in order to describe such interactions the field strength variables are not sufficient. We need the vector potentials which in classical electromagnetism play only an auxiliary role. The vector potentials are required for the Hamiltonian formalism necessary for quantization. The problem arises with the covariant quan-

© The Author(s), under exclusive license to Springer Nature Switzerland AG 2023   147
Z. Haba, *Lectures on Quantum Field Theory and Functional Integration*,
https://doi.org/10.1007/978-3-031-30712-6_9

tization of the four-potentials. Not all the components have a physical meaning. We must choose the correct number of degrees of freedom for quantization. We can distinguish physical components imposing a gauge condition. However, the gauge conditions which allow proper quantization (satisfying the condition of a positive scalar product in Hilbert space) violate the Lorentz invariance of vector potentials. The quantization with an indefinite metric in quantum electrodynamics (QED) which is Lorentz covariant but violates the positivity of the scalar product (in Hilbert space) has been developed by Gupta and Bleuler [37, 129, 214]. The task is to show that the physical correlation functions of field strength and scattering amplitudes do not depend on the method of quantization and the choice of gauge so that we can choose the gauge appropriate for the problem under consideration. As we show in Chap. 11 the gauge independence of the correlation functions of gauge invariant variables follows easily from the Fadeev-Popov formalism. In this chapter we work in various gauges and postpone the proof till Chap. 11 of the fact that the physics does not depend on the choice of gauge. We first quantize the electromagnetic field defining the functional integral by an introduction of a particular gauge fixing (the ξ-gauge, Sect. 9.1). The functional integral is Lorentz (or Euclidean) invariant. However, the corresponding Wightman functions (or Euclidean Schwinger functions) do not satisfy the positivity requirement. This positivity property is satisfied in the $A_0 = 0$ gauge. Hence, gauge invariant correlations will be positive definite. We define the Higgs model (Sect. 9.2) and represent the correlation functions of this model as a Gibbs ensemble of polymers (Sect. 9.3). Such a representation can be treated nonperturbatively at one-loop. In Sect. 9.4 we discuss the one-loop Euclidean effective action expressed by a determinant of the Laplacian in an external vector potential. In the subsequent sections we discuss a particle in a quantum electromagnetic field quantizing the field in the radiation gauge ($A_0 = 0$ and $\nabla \mathbf{A} = 0$). We calculate the transition amplitudes and the density matrix in an environment of photons in Sect. 9.5. In Sect. 9.6 we study the particle motion in quantum electromagnetic field working in the Heisenberg picture. If we do not observe photons then they play the role of a quantum noise. We derive a stochastic equation describing the radiation damping of the electron motion in the photon thermal state. The photon noise leads to the decoherence. This property is a consequence of the Lindblad equation satisfied by the particle density matrix (Sect. 9.9). The final section concerns the notion of entropy in quantum field theory. The functional integral approach leads to a definition of entropy similar to the one in classical random systems. We calculate the entropy of a Gaussian Wigner function.

9.1 Functional Integral Quantization of the Electromagnetic Field

In the description of electromagnetism a choice of units is relevant for a measurement of physical fields [118]. The various choices differ by a multiplication of fields by the electric ϵ_0 and magnetic μ_0 permeability of the vacuum. We choose the SI units. Then, the Lagrangian for a free electromagnetic field is [118]

$$\mathcal{L} = \frac{1}{2}(\epsilon_0 \mathbf{E}^2 - \mu_0^{-1}\mathbf{B}^2),$$

where $\mathbf{E}$ is the electric field, $\mathbf{B}$ the magnetic field and $\epsilon_0\mu_0 = c^{-2}$, where c is the velocity of light. We additionally choose the units such that $c = 1$ then we have a symmetry between $\mathbf{E}$ and $\mathbf{B}$. In such a case the Lagrangian for the free electromagnetic field can be expressed as

$$\mathcal{L} = -\frac{1}{4\mu_0} F_{\mu\nu} F^{\mu\nu}, \tag{9.1}$$

where the tensorial field strength is related to the electric and magnetic fields $F_{0j} = E_j$ and $B_j = \frac{1}{2}\epsilon_{jkl}F_{kl}$. In quantum theory we must express the field strength by vector potentials

$$F_{\mu\nu} = \partial_\mu A_\nu - \partial_\nu A_\mu. \tag{9.2}$$

The Lagrangian is invariant under the (gauge) transformation $A_\mu \rightarrow A_\mu + \partial_\mu f$. The action in terms of potentials is

$$\int dx\mathcal{L} = \frac{1}{2}\int dx A_\mu(-\eta^{\mu\nu}\partial_\alpha\partial^\alpha + \partial^\mu\partial^\nu)A_\nu. \tag{9.3}$$

We note that the differential operator defined by the rhs of Eq. (9.3) is not invertible because it is equal to zero if $A_\nu = \partial_\nu f$ as

$$(-\eta^{\mu\nu}\partial_\alpha\partial^\alpha + \partial^\mu\partial^\nu)\partial_\nu f = 0. \tag{9.4}$$

Equation (9.4) is a consequence of the gauge invariance. We encounter the difficulty that we cannot use our basic formula for a computation of correlation functions

$$\int dX \exp\left(-X\frac{M}{2}X\right)\exp(JX)\left(\int dX \exp\left(-X\frac{M}{2}X\right)\right)^{-1}$$
$$= \exp\left(\frac{1}{2}JM^{-1}J\right)$$

Physicists in the 50-ties invented a way out of the problem adding to the action the gauge fixing term

$$\frac{1}{2\xi}(\partial_\alpha A^\alpha)^2. \tag{9.5}$$

Now, the action can be written as

$$\int dx\mathcal{L} = \frac{1}{2\mu_0}\int dx A_\mu\left(-\eta^{\mu\nu}\partial_\alpha\partial^\alpha + \left(1 - \frac{1}{\xi}\right)\partial^\mu\partial^\nu\right)A_\nu. \tag{9.6}$$

The equation for the propagator $D^{F,\xi}$ reads

$$\left(-\eta^{\mu\nu}\partial_\alpha\partial^\alpha + \left(1-\frac{1}{\xi}\right)\partial^\mu\partial^\nu\right)D^{F,\xi}_{\mu\sigma}(x-y) = \mu_0\delta^\nu_\sigma\delta(x-y). \qquad (9.7)$$

$\xi = 0$ corresponds to the Lorentz gauge $\partial^\mu A_\mu = 0$, $\xi = 1$ to the Feynman gauge when the propagator is like the one for independent scalar fields (for a canonical quantization of the electromagnetic field in various ξ gauges, see [81]).

The solution of Eq. (9.7) is (in the momentum four-space)

$$D^{F,\xi}_{\mu\sigma}(p) = \mu_0(\eta_{\mu\sigma} - (1-\xi)p_\mu p_\sigma p^{-2})p^{-2}. \qquad (9.8)$$

The Euclidean continuation takes the form $D^{F,\xi}_{\mu\sigma}(p) \to D^{E,\xi}_{\mu\sigma}(p)$ where

$$D^{E,\xi}_{\mu\sigma}(p) = \mu_0(\delta_{\mu\sigma} - (1-\xi)p_\mu p_\sigma p_E^{-2})p_E^{-2}$$

with $p_E^2 = p_0^2 + \mathbf{p}^2$. There will be no reflection positivity of the generating functional

$$Z[f] = <\exp\left(i\int dx A_\mu f_\mu\right) >= \exp\left(-\frac{1}{2}f_\mu D^{E,\xi}_{\mu\sigma}f_\sigma\right)$$

as there is no quantum relativistic four-vector potential.

The gauge

$$A^0 = 0 \qquad (9.9)$$

in the Euclidean formulation renders OS positivity. This can be seen from the discussion of Sect. 3.10 because the spatial components of the vector potential $\mathbf{A}$ (from the point of view of the reflection positivity) can be treated like scalar fields (see Eq. (3.86)).

In the real time formulation, if equations of motion $\partial_\mu F^{\mu\nu} = 0$ are satisfied, then we can show that by means of the gauge transformation

$$A_\mu \to A_\mu + \partial_\mu f \qquad (9.10)$$

we can achieve $A^0 = 0$ and $\partial_k A^k = 0$. So that equations of motion read

$$\partial_\mu\partial^\mu A_k = 0, \qquad (9.11)$$

where $k = 1, 2, 3$. This is the radiation gauge applied in Sects. 9.5–9.10.

The question arises whether the various choices of gauge as (9.5) or (9.9) lead to equivalent theories. For the quantum field theory of the electromagnetic field the answer is confirmative as will be explained in Chap. 11 (in the context of general gauge theories when a modification of the functional integral is necessary). We show that gauge invariant correlation functions in theories (9.5) and (9.9) are the same. The gauge independence of scattering amplitudes in some gauges has been shown in

[33]. We may consider the renormalization of Euclidean gauge invariant correlation functions in the gauge (9.5) but in order to prove the OS positivity we can use the temporal gauge (9.9).

9.2 The Abelian Higgs Model

We are going to couple (in a gauge invariant way) the electromagnetic field to scalar fields. Assume we have two independent scalar fields ϕ_1 and ϕ_2. Consider the Lagrangian which is invariant under rotations

$$\phi_1' = \cos(\alpha)\phi_1 + \sin(\alpha)\phi_2, \tag{9.12}$$

$$\phi_2' = \cos(\alpha)\phi_2 - \sin(\alpha)\phi_1. \tag{9.13}$$

As shown in Sect. 2.1 there is a conserved current (electric current) resulting from this symmetry. As an invariant Lagrangian we take ($a = 1, 2$)

$$\mathcal{L} = \frac{1}{2}\partial_\mu\phi^a\partial^\mu\phi^a - \frac{1}{2}m^2\phi^a\phi^a - \frac{\lambda^2}{8}(\phi^a\phi^a)^2. \tag{9.14}$$

If we form

$$\phi = \phi_1 + i\phi_2, \tag{9.15}$$

then the rotation (9.12)–(9.13) can be expressed as the transformation $\phi \to \exp(-i\alpha)\phi$. The Lagrangian (9.14) takes the form

$$\mathcal{L} = \frac{1}{2}\partial_\mu\phi^*\partial^\mu\phi - \frac{m^2}{2}\phi^*\phi - \frac{\lambda^2}{8}(\phi^*\phi)^2. \tag{9.16}$$

We can couple the electromagnetic field to the scalar field in a way that preserves the gauge invariance

$$\phi \to \exp(-ief)\phi, \tag{9.17}$$

$$A_\mu \to A_\mu + \partial_\mu f. \tag{9.18}$$

The corresponding Lagrangian is (with an addition of the gauge fixing term $\frac{1}{2\xi}(\partial_\alpha A^\alpha)^2$)

$$\begin{aligned} \mathcal{L} = &-\frac{1}{4\mu_0}F_{\mu\nu}F^{\mu\nu} + \frac{1}{2\xi}(\partial_\alpha A^\alpha)^2 \\ &+\frac{1}{2}\big((\partial_\mu + ieA_\mu)\phi\big)^*\big((\partial^\mu + ieA^\mu)\phi\big) - \frac{1}{2}m^2\phi^*\phi - \frac{\lambda^2}{8}(\phi^*\phi)^2, \end{aligned} \tag{9.19}$$

where e is the electric charge. In the Higgs model $m^2 = -\mu^2 < 0$. Then, the scalar field gains a mass through a non-zero expectation value v of ϕ at the minimum $v = <\phi_1 >= \pm\frac{\sqrt{2}\mu}{\lambda}$ of the potential (9.19), so that under the shift $\phi_1 = \tilde{\phi}_1 + v$ (where $\tilde{\phi}$ has zero mean value) we obtain from the quadratic part of the Lagrangian (9.19) that the field $\tilde{\phi}_1$ has the mass μ and the electromagnetic field the mass $e|v|\sqrt{\mu_0}$.

We restrict our discussion to large ϕ when the mechanism of gaining the mass does not play a substantial role (for small values of ϕ we should consider the bilinear form in $\tilde{\phi}$ rather than in ϕ). The quantum theory is defined by the functional integral

$$\int d\phi dA \exp\left(\frac{i}{\hbar}\int dx(\mathcal{L}_0 + \mathcal{L}_I)\right), \tag{9.20}$$

where $\mathcal{L}_0$ is the quadratic part of $\mathcal{L}$ and

$$\mathcal{L}_I = -\frac{\lambda^2}{8}(\phi^*\phi)^2 + \frac{1}{2}e^2 A_\mu A^\mu \phi^*\phi + \frac{1}{2}ieA^\mu(\partial_\mu\phi^*\phi - \partial_\mu\phi\phi^*). \tag{9.21}$$

When we expand the exponential of $\mathcal{L}_I$ in Eq. (9.20) in the power series in the couplings λ and e, then we obtain polynomials in ϕ, ϕ^*, A_μ. Expectation values of polynomials are expressed by means of propagators. Only $\phi^*\phi$ correlations (with an equal number of ϕ and ϕ^*) are different from zero (because of the invariance (9.17)). They are expressed by the propagator $i\Delta_F$ (the propagator for $\partial_\mu\phi^*(x)\phi(y)$ is $\partial_\mu i\Delta_F(x-y)$). The propagator of the electromagnetic field is

$$D_{\mu\sigma}^{F,\xi}(x-y) = \mu_0(2\pi)^{-4}\int dp\exp(-ip(x-y))(\eta_{\mu\sigma} - (1-\xi)p_\mu p_\sigma p^{-2})p^{-2} \tag{9.22}$$

(for a discussion of the integration over p in Eq. (9.22), see [81]; there is an extra p^{-2} term in comparison to the scalar propagator discussed at the end of Sect. 3.5).

In the lowest order in λ the integral over ϕ (at $\lambda = 0$) can be calculated explicitly. In this approximation the generating functional is

$$\begin{aligned}
Z[J, j] &=< \exp(\frac{i}{\hbar}(\int J^*\phi + \int J\phi^* + \int j_\mu A_\mu) > \\
&= \int dA \exp(-\frac{i}{4\hbar\mu_0}\int F^2)\exp\left(i\int j_\mu A_\mu\right) \\
&\times \exp\left(-\frac{i}{\hbar}\int J^*\left(-(\partial_\mu + ieA_\mu)^+(\partial^\mu + ieA^\mu) + m^2\right)^{-1}J\right) \\
&\times \det\left(-(\partial_\mu + ieA_\mu)^+(\partial^\mu + ieA^\mu) + m^2\right)^{-\frac{1}{2}}.
\end{aligned}$$

For the electromagnetic field we obtain a non-linear electrodynamics with the functional integral of the form

$$\int dA \exp\left(-\frac{i}{4\hbar\mu_0}\int dx F^2\right)\det\left(-(\partial_\mu + ieA_\mu)^+(\partial^\mu + ieA^\mu) + m^2\right)^{-\frac{1}{2}}. \tag{9.23}$$

In Euclidean version we shall have in Eq. (9.23) the determinant of the operator

$$M_A = -(\partial_\mu + ieA_\mu)^+(\partial_\mu + ieA_\mu) + m^2, \tag{9.24}$$

$\mu = 0, 1, 2, 3.$

9.3 Euclidean Version: The Polymer Representation

In this section we apply the Feynman integral of Chap. 5 in order to represent the Higgs model as a sum of Wiener paths with a certain Gibbs factor describing an interaction between paths. For this purpose let us first introduce an auxiliary real field Φ by means of the formula

$$\exp\left(-\frac{\lambda^2}{8}\int dx(\phi^*\phi)^2\right) = \int d\Phi \exp\left(-\frac{1}{2}\int dx\Phi^2\right)\exp\left(\frac{i\lambda}{2}\int dx\Phi\phi^*\phi\right). \tag{9.25}$$

With the representation (9.25) the propagator for the ϕ field in external A_μ and Φ fields is (we set $\hbar = 1$ and we choose the Lorentz gauge $\xi = 0$, then the Ito and Stratonovitch integrals are equal)

$$M^{-1}(x, y) = \left(-(\partial_\mu + ieA_\mu)^+(\partial_\mu + ieA_\mu) + m^2 + i\lambda\Phi\right)^{-1}(x, y)$$
$$= \int_0^\infty dt\, dW^t_{(x,y)}(w)\exp(-\tfrac{1}{2}m^2 t)\exp\left(ie\int_0^t A_\mu(w)dw_\mu + i\lambda\int_0^t \Phi(w(s))ds\right), \tag{9.26}$$

where we used the representation $M^{-1} = \int_0^\infty dt\exp(-tM)$ and the formula from Sect. 5.4 for the representation of the operator $\exp(-tM)$ $(M = M_A + i\lambda\Phi)$. All the correlation functions in the Higgs model can be expressed by Gaussian integrals over A_μ, Φ and w. For example

$$< \phi^*(x)\phi(y) >= \left\langle M^{-1}(x, y)\det M^{-\frac{1}{2}}\right\rangle, \tag{9.27}$$

where $\det M$ can also be expressed by the Wiener integral according to Eqs. (7.62)-(7.64) ($i\lambda\Phi$ is like a potential). Then, the expectation value (9.27) is over the fields Φ and A_μ. The determinant in the representation (7.62) is an exponential function. When we expand the exponential then we obtain a Gaussian integral over Φ and A_μ. In the zeroth order of the expansion of the exponential (i.e.,neglecting the determinant) the expectation value (9.27) is $(x, y \in R^4)$

$$< \phi^*(x)\phi(y) >= \int_0^\infty dt\exp(-\tfrac{1}{2}m^2 t)dW^t_{(x,y)}(w)\exp\left(-\frac{e^2}{2}\int_0^t\int_0^t\right.$$
$$\times D^{\mu\nu}_{E,0}(w(s) - w(s'))dw^\mu(s)dw^\nu(s') - \tfrac{\lambda^2}{2}\int_0^t\int_0^t \delta(w(s) - w(s'))dsds'\Big), \tag{9.28}$$

because the covariance of $\Phi(x)$ is $\delta(x - x')$. The last term in the exponential (9.28) looks more singular then the term with the stochastic integral but we can transform it to another form using the formula (5.60)

$$df = \partial_\mu f dw_\mu + \frac{1}{2}\partial_\mu\partial_\mu f ds. \tag{9.29}$$

We set $\Phi = \Delta_E f$, where $\Delta_E = \partial_\sigma\partial_\sigma$, in Eq. (9.26) and we rewrite Eq. (9.29) in the integral form

$$\int_0^t \frac{1}{2}\Phi(w(s))ds = -\int_0^t \partial^\mu\Delta_E^{-1}\Phi dw^\mu + \Delta_E^{-1}(\Phi(w(t)) - \Phi(x)), \tag{9.30}$$

where $w(t) = y$ in Eq. (9.30). Then, the δ term in the exponential (9.28) can be expressed by three terms: the first is similar to the electromagnetic term

$$- 2\lambda^2 \int_0^t \int_0^t G_{\mu\nu}(w(s) - w(s'))dw_\mu(s)dw_\nu(s'), \tag{9.31}$$

where

$$G_{\mu\nu}(x - y) = \partial_\mu\partial_\nu\Delta_E^{-2}(x - y).$$

Both $D_{\mu\nu}^{E,0}$ and $G_{\mu\nu}$ have the same singularity at short distances. Moreover, the G term can be interpreted as a shift of the ξ parameter in the ξ-gauge (9.22). The second term which will appear in (9.28) comes from

$$\int d\Phi \exp(-\tfrac{1}{2}\int dx\Phi^2 + 2i\lambda\Delta_E^{-1}\Phi(y) - 2i\lambda\Delta_E^{-1}\Phi(x))$$
$$= \exp\left(4\lambda^2 \int dz\Big(\Delta_E^{-1}(x - z)\Delta_E^{-1}(y - z)\right)$$
$$-\tfrac{1}{2}\Delta_E^{-1}(x - z)\Delta_E^{-1}(x - z)) - \tfrac{1}{2}\Delta_E^{-1}(y - z)\Delta_E^{-1}(y - z)\Big)\Big).$$

It is vanishing when $x \to y$. The third term is the cross term from the integration over Φ of the rhs of Eq. (9.30). It has the form of a stochastic integral

$$\int_0^t dw^\mu(s) \int du \partial^\mu\Delta_E^{-1}(w(s) - u)(\Delta_E^{-1}(u - y) - \Delta_E^{-1}(u - x)). \tag{9.32}$$

The integral (9.32) is a singular stochastic integral, but in the limit $x \to y$ we get zero. The representation (9.28)–(9.32) (without the electromagnetic field, for the electromagnetic field see [14]) has been discussed first by Symanzik [229]. In [229] it is proved that in $d = 2$ after a renormalization of the δ-function the expression

(9.28) is finite (with a finite range of the t-integral in Eq. (9.28)). The construction of the $d = 2$ Higgs model has been achieved by other means in [43].

The determinant is represented in terms of paths as

$$
(\det M)^{-\frac{1}{2}} = \exp\left(\frac{1}{2}\int dx \int_0^\infty \frac{dt}{t} dW_{(x,x)}^t(w)\exp(-\tfrac{1}{2}m^2 t)\right)
$$
$$
\left(\exp\left(ie\int_0^t A_\mu(w)dw_\mu + i\lambda \int_0^t \Phi(w(s))ds\right) - 1\right) \tag{9.33}
$$
$$
\equiv \exp(-\Gamma) = \sum_{n=0}^\infty \frac{1}{n!}(-\Gamma)^n.
$$

After the expansion (9.33) we can perform the Gaussian integrals over A_μ and Φ. As a result we obtain exponentials of the form

$$
\exp\left(-\sum_{j,k}\int_0^{t_j}\int_0^{t_k}(\tfrac{e^2}{2}D_{E,0}^{\mu\nu} + 2\lambda^2 G^{\mu\nu})(w_j(s) - w_k(s'))dw_j^\mu(s)dw_k^\nu(s')\right),
$$

where w_j are closed Brownian paths and

$$
\exp\left(\sum_j \int_0^t \int_0^t (e^2 D_{E,0}^{\mu\nu} + 4\lambda^2 G^{\mu\nu})(w_j(s) - w(s'))dw_j^\mu(s)dw^\nu(s')\right.
$$
$$
\left. - \sum_j \int_0^t \int_0^t (\tfrac{e^2}{2}D_{E,0}^{\mu\nu} + 2\lambda^2 G^{\mu\nu})(w(s) - w(s'))dw^\mu(s)dw^\nu(s')\right),
$$

where w_j are closed Brownian paths and w is the Brownian motion starting from x and ending in y. The expression for the correlation functions of scalar fields looks like an expectation value of polymer interaction terms in a grand canonical ensemble summed over the canonical Gibbs ensembles of n-paths. For the partition function in the model (9.19) (equal to $\det M^{-\frac{1}{2}}$) we would obtain exactly the grand partition function summed over the partition functions of n interacting polymers. Each term in such an expansion can be treated as a statistical Gibbs ensemble of paths interacting through the potential $D^{E,\xi}(x - y)$ (for some studies of ensembles of polymer paths and the corresponding stochastic integrals see [123, 170, 229, 241], the double stochastic integrals are studied in [51]; the first paper suggesting the use of paths instead of potentials was that of Mandelstam [182]). The expansion (9.33) involves singular stochastic integrals. It needs a regularization and renormalization. We shall discuss the lattice version of the polymer representation in Chap. 12. Then, the Wiener paths are replaced by the Poissonian paths which move on the sites of the lattice. The replacement results from the definition of gauge and scalar fields which live on the lattice. Using the polymer representation some results on the mass gap in the ϕ^4 model have been established [44, 45] (independent of the lattice spacing). In [95] it has been proved (using the polymer representation) that there is no non-trivial ϕ^4 continuum limit in more than four dimensions.

In Eq. (9.33) from the contribution of the determinant we have closed paths whereas from the scalar fields in the correlation functions the paths are open. We can change this inconvenient statistical picture if we introduce a gauge invariant variable

$$\int_0^\infty dt \exp\left(-\frac{1}{2}m^2 t\right) dW_{(y,x)}^t(w)\phi^*(y) \exp\left(i \int_0^t A_\mu(w(s)) \circ dw_\mu(s)\right)\phi(x) \quad (9.34)$$

When computing correlation functions of gauge invariant variables (9.34) the integration over Brownian paths in Eq. (9.28) can be expressed by closed paths. For such a representation the following identity is useful: let $dW_{(x,y)}^{(s,t)}$ be the Wiener measure on paths such that $w(s) = x$ and $w(t) = y$. Then, for $s \le t \le \tau$

$$\int dy dW_{(x,y)}^{(s,t)}(w) dW_{(y,z)}^{(t,\tau)}(w') = dW_{(x,z)}^{(s,\tau)}(w'') \quad (9.35)$$

in the sense that for $s \le t_1 \le \ldots \le t_k \le t$ and $t \le t_{k+1} \le \ldots \le t_n \le \tau$ and an arbitrary function F we have

$$\begin{aligned}
&\int dy dW_{(x,y)}^{(s,t)}(w) dW_{(y,z)}^{(t,\tau)}(w') F(w(t_1), \ldots, w(t_k), w'(t_{k+1}), ..w'(t_n)) \\
&= \int dW_{(x,z)}^{(s,\tau)}(w'') F(w''(t_1), \ldots, w''(t_k), w''(t_{k+1})...w''(t_n)).
\end{aligned} \quad (9.36)$$

Another approach to gauge invariant polymer quantization of the Higgs model, which as in Eq. (9.34) works solely with closed paths, would involve an introduction of the string fields

$$\exp\left(i \int_0^t A_\mu(\xi^x(s)) d\xi_\mu^x(s)\right)\phi(x),$$

where ξ^x is a path from $-\infty$ to x. Then, we consider the generating functional of the string fields

$$\Big\langle \exp\left(i \int dx J(x) \exp(i \int_0^t A_\mu^x(\xi(s)) d\xi_\mu^x(s))\phi(x)\right. \\
\left. + \int dx J^*(x) \exp(-i \int_0^t A_\mu^x(\xi(s)) d\xi_\mu^x(s))\phi^*(x)\right) \Big\rangle$$

which can be calculated by an expansion in J (see [23]).

There are still the correlation functions of gauge potentials. Instead of local functions of gauge potentials we should rather use the gauge invariant closed loop integrals (Wilson loops)

$$\exp\left(ie \int A_\mu dq^\mu\right),$$

where q^μ is a closed curve. The Gaussian integral over the electromagnetic field is then expressed by

$$\exp\left(-\frac{e^2}{2} \int dq^\mu(s) dq^\nu(s') D_{E,\xi}^{\mu\nu}(q(s) - q(s'))\right)$$

The Euclidean Higgs model as well as the related polymer models are interesting for quantum field theory as well as for statistical physics (Ginzburg-Landau model) in dimension $d \leq 3$. The polymer representation began with the Symanzik paper [229]. For more recent studies see [14, 45, 83, 95, 212] (the last reference shows a relation between the polymer representation and string theory).

9.4   One-Loop Determinant: A Non-Perturbative Method

The one-loop contribution to the Higgs model is expressed by the determinant of the operator (M_A is defined in Eq. (9.24))

$$M = M_A + 24\lambda|\phi_c|^2 \equiv M_A + V. \tag{9.37}$$

We can allow in Eq. (9.37) $m^2 < 0$ if $m^2 + 12\lambda|\phi_c|^2 > 0$. In order to calculate the determinant we apply the methods of Sect. 7.6

$$\det M = \exp(Tr \ln M) = \exp\left(-Tr \int\limits_0^\infty \frac{dt}{t} \exp(-tM)\right). \tag{9.38}$$

We need to investigate the operator $\exp(-tM)$. We apply the Feynman formula (Sect. 5.4) in the Euclidean framework ($x, y \in R^d$)

$$\exp(-tM)(x, y) = \int dW_{(x,y)}^t(w) \exp\left(ie \int_0^t A_\mu \circ dw_\mu - \int_0^t V(w(s))ds\right)$$

$$\equiv \int dW_{(x,y)}^t(w) T_t^A \exp\left(-\int_0^t V(w(s))ds\right),$$

where $V = m^2 + 12\lambda|\phi_c|^2$ and $T_t^A = \exp(ie \int_0^t A_\mu \circ dw_\mu)$.

The determinant describes the one loop contributions to the effective action. We expect (as in Sects. 7.5 and 7.6) that the ultraviolet divergencies will show up and will be cured by renormalization. In the determinant there appears

$$\int\limits_0^\infty \frac{dt}{t} \int dW_{(x,x)}^t(q)\left(T_t^A \exp(-\int_0^t V(q(s))ds) - 1\right). \tag{9.39}$$

We changed the notation $w \to q$ for the closed paths w in order to apply the representation (7.65) and (7.67)

$$q(s) = x + \sqrt{t}r\left(\frac{s}{t}\right) = x + \sqrt{t}\left(1 - \frac{s}{t}\right)w\left(\frac{\frac{s}{t}}{1 - \frac{s}{t}}\right),$$

where w is unconstrained Brownian motion with the covariance (5.31) and $\int dW_{(x,x)}^t(q)f(q) = (2\pi t)^{-\frac{d}{2}} E[f]$ (see Eq. (5.35)). Equation (9.39) is divergent at small t. Such an expression with $A = 0$ and $V \neq 0$ has been studied in Sect. 7.6. In this section we restrict ourselves to $V = 0$ and $A \neq 0$.

In order to determine the divergence we expand A_μ in terms of the derivatives of the electromagnetic field. For this purpose it is useful to apply the coordinate gauge for each $x \in R^d$ [73]

$$y_\mu A_\mu(x + y) = 0. \tag{9.40}$$

In the coordinate gauge we can express the gauge potential in terms of the field strength

$$A_\mu(x + y) = \int_0^1 d\alpha F_{\nu\mu}(x + \alpha y)\alpha y_\nu = \frac{1}{2}F_{\nu\mu}(x)y_\nu + C_\mu(\partial F) \equiv B_\mu + C_\mu,$$
$$\tag{9.41}$$

where C_μ can be expressed by $\partial_\alpha F_{\nu\mu} y^\nu y^\alpha$. In Eq. (9.39) $y = \sqrt{t} tr(\frac{s}{t})$. We write

$$T_t^{B+C} - 1 = (T_t^B - 1) + T_t^B(T_t^C - 1). \tag{9.42}$$

It can be seen that the $F_{\mu\nu}$ term in (9.41)–(9.42) leads to a logarithmic divergent term $\int dt t^{-1} F_{\mu\nu} F_{\mu\nu}$ in Eq. (9.39) because $E[(T^B - 1)] \simeq E[t^2 \int (Frdr)^2]$ for a small t. After a subtraction of this term from (9.39) the remaining expectation value will be of order t^3. Hence, the $(T^B - 1)$ term in (9.42) after the renormalization when inserted in (9.39) will be already convergent at small t. As $Cdq \simeq \sqrt{t}(\sqrt{t}r)^2$ the average with respect to the Wiener measure over the second term in the decomposition (9.42) behaves as $(\int Cdq)^2 \simeq t^3$ for a small t (hence it has a convergent t-integral in Eq. (9.39)) and is bounded by a quadratic functional of ∂F [134]. It follows that the behavior of the determinant depends in a crucial way on the first term in the decomposition (9.41)–(9.42) which has the form of the integral

$$\exp(\frac{t}{2}\Delta_A)(x, x) = \int dW_{(x,x)}^t(w) \exp\left(\frac{ie}{2} \int_0^t F_{\mu\nu}(x)w_\mu dw_\nu\right). \tag{9.43}$$

The determinant defines the effective action at one-loop. We are interested in an exact calculation of this effective action in $d = 4$. The problem involves a calculation of the Wiener integral (9.43). This integral is expressing the imaginary time evolution kernel in quantum mechanics with a constant field strength $F_{\mu\nu}$ in d dimensions. One can reduce the problem of the calculation of the integral (9.43) to the one appearing in the formula for the harmonic oscillator (Mehler formula) [135, 224].

Another method based on the Heisenberg picture in quantum mechanics has been invented by Schwinger [215]. In this method we first solve the Heisenberg equations of motion

$$\frac{dx^\mu}{ds} = 2\Pi^\mu,$$ (9.44)

$$\frac{d\Pi_\mu}{ds} = 2e F_{\mu\nu} \Pi_\nu$$ (9.45)

for the Hamiltonian

$$H = \Pi^\mu \Pi^\mu.$$ (9.46)

Equations (9.44)–(9.45) can be treated as matrix equations (F is a matrix with matrix elements $F_{\mu\nu}$). In order to calculate (9.43) we need to compute

$$< x| \exp(-sH)|y >$$ (9.47)

using the expression of H in terms of $x(s)$ and $x(0)$. The result is

$$Tr < x| \exp(-sH)|x >= (4\pi)^{-2} \exp(-L(s)),$$ (9.48)

where

$$L(s) = \frac{1}{2} Tr \ln[(eFs)^{-1} \sinh(eFs)].$$ (9.49)

There remains to evaluate the trace (9.49) (over the $\mu\nu$-indices). It can be expressed by two scalars

$$\mathcal{F} = e^2 F_{\mu\nu} F_{\mu\nu}$$ (9.50)

and

$$\mathcal{G} = -\frac{e^2}{2} F_{\mu\nu} \epsilon_{\mu\nu\sigma\rho} F_{\sigma\rho}.$$

Let

$$u_\pm = \frac{\sqrt{2}}{8}\left(\sqrt{\mathcal{F}+\mathcal{G}} \pm \sqrt{\mathcal{F}-\mathcal{G}}\right).$$ (9.51)

Then

$$\exp\left(\frac{s}{2}\Delta_A\right)(x, x) = (2\pi s)^{-2}\left(1 - u_+ u_- (\sinh(u_+ s) \sinh(u_- s))^{-1}\right).$$ (9.52)

The kernel (9.52) allows to calculate the determinant according to Eq. (7.62). When we insert the kernel (9.52) in the formula (9.39) then we notice that it requires a subtraction of the term (encountered already at Eq. (9.42))

$$\int_0^\infty \frac{dt}{t} \exp\left(-\frac{1}{2}m^2 t\right)\mathcal{F}$$ (9.53)

which can be interpreted as the charge renormalization of the initial action

$$\frac{1}{\mu_0 e_{bare}^2}\mathcal{F} = \frac{1}{\mu_0 e^2}\mathcal{F} + K \int\limits_0^\infty \frac{dt}{t} \exp\left(-\frac{1}{2}m^2 t\right)\mathcal{F},$$

where $K < 0$. Unfortunately, $e_{bare}^2 < 0$ when $K < 0$. In the effective action of the non-Abelian gauge theories we obtain $K > 0$, see Chap. 11.

Using the decomposition (9.41)–(9.42) and the result (9.52) one can derive an estimate on the determinant

$$\det M^{-\frac{1}{2}} = \exp\left(\frac{1}{48} \int dx \mathcal{F} \ln \mathcal{F} + Q(F, \partial F)\right),$$

where $\mathcal{F} \simeq F^2$ and $Q(F, \partial F)$ is quadratically bounded in F and ∂F. It follows [134] that $\det M^{-\frac{1}{2}}$ is not integrable with respect to the Gaussian measure of the electromagnetic field and the scalar QED is unstable. It can be seen from the derivation that the instability is a consequence of the charge renormalization (9.53).

The effective action (in the approximation of constant electric and magnetic fields) has been calculated a long time ago by Euler and Heisenberg [77] (in Minkowski space). It allows to investigate classically non-linear effects (non-linear electrodynamics) resulting from quantum phenomena. For recent studies of the determinant in quantum electrodynamics (QED) and in the Standard Model see [97]. We shall continue the investigation of the effective action resulting from an interaction of non-Abelian gauge fields with a scalar field in Chap. 11.

9.5 Non-relativistic QED: A Charged Particle Interacting with Quantum Electromagnetic Field

We discuss in this section the well-known quantum-mechanical model [75, 151, 188] of a non-relativistic electron in a quantum electromagnetic field. A straightforward approach applies the Heisenberg equations of motion. In such a description the creation and annihilation operators of the electromagnetic field play the role of the quantum noise (quantum Brownian motion as discussed in [5, 6, 30]). General properties of solutions have been obtained in this approach [91, 92]. The method is fruitful in applications to quantum optics [101]. Some basic results can be obtained already in the classical approximation (e.g., the Abraham-Lorentz radiation damping responsible for the spectral line width [151]). In a general non-linear case it is difficult to solve Heisenberg equations of motion and calculate correlation functions (beyond the perturbative expansion in the coupling constant). Another approach is using the functional integral method (the influence functional [86]). In such a formulation one can obtain, in a semiclassical expansion, a classical stochastic equation which

determines correlation functions, scattering probabilities and the density matrix of a
particle moving in an electromagnetic environment (see [48, 86, 142, 156]).

In quantum electrodynamics we are interested in electron transitions involving
the photon emission. These phenomena can be studied by means of an expansion
in the electric charge e(more precisely in $\frac{e^2}{c\hbar}$, see [151, 188] Appendix D; [181]
Sect. 14). The electron-photon interaction is studied in quantum optics experiments
in terms of the photometric detection [152, 181, 188]. In some recent papers [38,
167, 193, 194] a related question has been raised of whether we can detect gravitons
on the basis of particle-graviton interaction. In particular, in [193, 194] it has been
suggested that we can look for the gravitons through the study of the quantum noise
produced by these particles. The Heisenberg equations of motion for such a particle-
graviton interaction have been investigated in [145, 167] from this point of view.
In this chapter we study the analogous problem in quantum electrodynamics. The
graviton-particle interaction is discussed in the next chapter.

We quantize canonically the free electromagnetic field in the radiation gauge.
Then, the Hamiltonian for a field-particle interaction is implementing the particle
time evolution. From the Lagrangian (9.1) $\partial_t \mathbf{A} = \mu_0 \Pi$. Hence, the Hamiltonian for
the electromagnetic field is ($\mathbf{B} = \nabla \times \mathbf{A}$)

$$H_0 = \tfrac{1}{2} \int d\mathbf{x}(-\hbar^2 \mu_0 \frac{\delta}{\delta \mathbf{A}(\mathbf{x})} \frac{\delta}{\delta \mathbf{A}(\mathbf{x})} + \mu_0^{-1} \mathbf{B}^2). \tag{9.54}$$

The ground state of H_0 is

$$\psi^g = Z^{-1} \exp\left(-\frac{1}{2\mu_0 \hbar}(\mathbf{A}, \sqrt{-\Delta}\mathbf{A})\right). \tag{9.55}$$

We apply the formalism of Sect. 6.3. The stochastic equation describing the electromagnetic field evolution in the ground state according to Chap. 6 has the form

$$d\mathbf{A} = -i\sqrt{-\Delta}\mathbf{A}dt + \sqrt{i\hbar\mu_0}\Lambda d\mathbf{W}, \tag{9.56}$$

where

$$\Lambda^{jl}(\nabla(-\Delta)^{-\frac{1}{2}}) = \delta^{jl} - \partial^j \partial^l \Delta^{-1}$$

is imposing the radiation gauge.

In the non-relativistic approximation the interaction of the electromagnetic field
with a charged particle described by a coordinate $\mathbf{q}$ has the Lagrangian (we set the
electric charge $e = 1$ and the velocity of light $c = 1$)

$$\mathcal{L} = \mathcal{L}_\mathbf{q} + \frac{1}{2\mu_0} \int d\mathbf{x}((\partial_s \mathbf{A})^2 - \mathbf{B}^2) + \mathbf{A}(\mathbf{q}(s))\frac{d\mathbf{q}}{ds}, \tag{9.57}$$

where $\mathcal{L}_\mathbf{q}$ is the particle Lagrangian.

We are interested in the scattering amplitude a_{fi} from an initial state which is a
product state of the state for the quantum electromagnetic field $\psi^g(\mathbf{A})\chi_i(\mathbf{A})$ (where

ψ^g is the ground state (9.55)) and the particle state ϕ_i. We study the transition from the initial state at t_0 $\psi^g(\mathbf{A})\chi_i(\mathbf{A})\phi_i(\mathbf{x})$ to the final state $\psi^g(\mathbf{A})\chi_f(\mathbf{A})\phi_f(\mathbf{x})$. We apply the transformation of Sect. 6.3 only to the electromagnetic path integral. Then, according to Eq. (8.6) the transition amplitude a_{fi} is (we denote $q_s = q(s)$)

$$
\begin{aligned}
a_{fi} &= (\psi^g \chi_f \phi_f, U(t, t_0)\psi^g \chi_i \phi_i) \\
&= \int d\mathbf{x} d\mathbf{A} \mathcal{D}\mathbf{q} \exp(\tfrac{i}{\hbar} \int ds \mathcal{L}_\mathbf{q})(\psi^g(\mathbf{A})\chi_f(\mathbf{A})\phi_f(\mathbf{x}))^* \\
&\quad \psi^g(\mathbf{A})\phi_i(\mathbf{q}_t(\mathbf{x}))E\left[\exp\left(\tfrac{i}{\hbar} \int_{t_0}^{t} ds \mathbf{A}(\mathbf{q}_s)\mathbf{v}_{t-s} \right)\chi_i(\mathbf{A}_t(\mathbf{A})) \right],
\end{aligned}
\tag{9.58}
$$

where $U(t, t_0)$ is the unitary evolution in the model (9.57), $\mathcal{D}\mathbf{q}$ denotes the Feynman integral over the paths of the system $\mathcal{L}_\mathbf{q}$, E[...] is the expectation value with respect to the stochastic electromagnetic field determined by Eq. (9.56) and

$$
v_s^j = \frac{dq^j}{ds} \tag{9.59}
$$

comes from the interaction in Eq. (9.57).

The initial state of the electromagnetic field is $\psi^g \chi_i$. If $\chi_i = 1$ (the initial state is the electromagnetic vacuum) then the expectation value in Eq. (9.58) can easily be calculated (here $< \mathbf{A}_s >= E[\mathbf{A}_s]$)

$$
\begin{aligned}
E\left[\exp\left(\tfrac{i}{\hbar} \int_{t_0}^{t} ds \mathbf{A}_s \mathbf{v}_{t-s} \right) \right] &= \exp\left(\tfrac{i}{\hbar} \int_{t_0}^{t} < \mathbf{A}_s > \mathbf{v}_{t-s} \right) \\
\exp\left(- \tfrac{1}{2\hbar^2} \int_{t_0}^{t} dsds' E[(A_s^j - < A_s^j >)(A_{s'}^l - < A_{s'}^l >)]v_{t-s}^j v_{t-s'}^l \right).
\end{aligned}
\tag{9.60}
$$

If $t_0 = -\infty$ then

$$
\begin{aligned}
E[(A_s^j - &< A_s^j >)(\mathbf{x})(A_{s'}^l - < A_{s'}^l >)(\mathbf{x}')] \\
&= (2\pi)^{-3} \int d\mathbf{p}(2|\mathbf{p}|)^{-1}\Lambda^{jl}(\mathbf{p}|\mathbf{p}|^{-1}) \exp(i(\mathbf{p}(\mathbf{x} - \mathbf{x}')) \exp(-i|\mathbf{p}||s - s'|) \\
&= i\Lambda^{jl}(\nabla(-\Delta)^{-\frac{1}{2}})\Delta_F(x - x'),
\end{aligned}
\tag{9.61}
$$

where on the rhs we have the time-ordered expectation value (propagator) of the vector potential in the radiation gauge

$$
\begin{aligned}
i\Delta_F(x - x') \\
= (2\pi)^{-3} \int d\mathbf{p}(2|\mathbf{p}|)^{-1} \exp(-i|\mathbf{p}||s - s'| - i\mathbf{p}(\mathbf{x} - \mathbf{x}')).
\end{aligned}
$$

We encounter in Eq. (9.60) the polymer interaction of Sect. 9.3 (but now in the radiation gauge).

If we do not observe the final states f of the electromagnetic field and average the probability $P(i, f)$ of the transition $\phi_i \rightarrow \phi_f$ over these states using the completeness relation

$$\sum_f (\psi^g(\mathbf{A})\chi_f(\mathbf{A}))^* \psi^g(\mathbf{A}')\chi_f(\mathbf{A}') = \delta(\mathbf{A} - \mathbf{A}') \qquad (9.62)$$

then we obtain

$$
\begin{aligned}
P(i \to f) &= \sum_f |a_{fi}|^2 \\
&= \int d\mathbf{A} \mathcal{D}\mathbf{q} \mathcal{D}\mathbf{q}' d\mathbf{x} d\mathbf{x}' |\psi^g(\mathbf{A})|^2 \phi_f^*(\mathbf{x})\phi_f(\mathbf{x}')\phi_i(\mathbf{q}_t(\mathbf{x}))\phi_i^*(\mathbf{q}_t'(\mathbf{x}')) \\
&\quad \exp(\tfrac{i}{\hbar}\int ds \mathcal{L}_{\mathbf{q}} - \tfrac{i}{\hbar}\int ds \mathcal{L}_{\mathbf{q}'}) E\Big[\exp\Big(\tfrac{i}{\hbar}\int_{t_0}^t ds \mathbf{A}_s \mathbf{v}_{t-s}\Big)\chi_i(\mathbf{A}_t(\mathbf{A})) \\
&\quad \exp\Big(-\tfrac{i}{\hbar}\int_{t_0}^t ds \mathbf{A}_s^* \mathbf{v}_{t-s}'\Big)\chi_i^*(\mathbf{A}_t^*(\mathbf{A})) \Big],
\end{aligned}
\qquad (9.63)
$$

where $\mathbf{A}_t^*$ denotes a complex conjugation of an independent version of the process $\mathbf{A}_t$ and $\mathbf{v}' = \frac{d\mathbf{q}'}{dt}$ is an independent version of the process $\frac{d\mathbf{q}_t}{dt}$, $\mathbf{x}, \mathbf{x}'$ are the initial values of the paths $\mathbf{q}$ (resp.$\mathbf{q}'$). The δ-function in Eq. (9.62) can be understood as the one defined in the Hilbert space $L^2(d\mu_0)$ with the Gaussian measure $d\mu_0 = |\psi^g|^2 d\mathbf{A}$. This corresponds to $\rho = \psi^g$ in the definition of the kernel in Eq. (3.50). Then, Eq. (9.62) is just a realization of Eq. (3.46) in the Hilbert space $L^2(d\mu_0)$ of functionals in an infinite number of dimensions.

On the other hand if we average the particle state $\rho = E[|i><i|]$ over the environment of the electromagnetic field then the particle density matrix is

$$
\begin{aligned}
\rho_t(\mathbf{x}, \mathbf{x}') &= \int d\mathbf{A} \mathcal{D}\mathbf{q} \mathcal{D}\mathbf{q}' |\psi^g(\mathbf{A})|^2 \exp(\tfrac{i}{\hbar}\int ds \mathcal{L}_{\mathbf{q}} - \tfrac{i}{\hbar}\int ds \mathcal{L}_{\mathbf{q}'})\phi_i(\mathbf{q}_t(\mathbf{x}))\phi_i^*(\mathbf{q}_t'(\mathbf{x}')) \\
&\quad E\Big[\exp\Big(\tfrac{i}{\hbar}\int_{t_0}^t ds \mathbf{A}_s \mathbf{v}_{t-s}\Big)\chi_i(\mathbf{A}_t(\mathbf{A})) \exp\Big(-\tfrac{i}{\hbar}\int_{t_0}^t ds \mathbf{A}_s^* \mathbf{v}_{t-s}'\Big)\chi_i^*(\mathbf{A}_t^*(\mathbf{A})) \Big].
\end{aligned}
\qquad (9.64)
$$

Comparing Eqs. (9.63) with (9.64) we obtain

$$P(i \to f) = \int \rho_t(\mathbf{x}, \mathbf{x}')\phi_f^*(\mathbf{x})\phi_f(\mathbf{x}') d\mathbf{x} d\mathbf{x}'.$$

We cannot calculate the transition probability exactly but we can do this in an expansion in $\hbar$ or in $\frac{e^2}{\hbar c}$ (there is $e = c = 1$ in Eq. (9.63)). The expansion in $\frac{e^2}{\hbar c}$ (which is an expansion in powers of the electromagnetic field) coincides with the standard calculations of photo detection done in [188] (Appendix E) or [181] (Sect. 14). Mathematical problems concerning regularization and renormalization of the formula (9.63) are discussed in [140, 179].

An evaluation of the average over $\mathbf{A}$ in Eq. (9.64) gives an effective action $\mathcal{L}(\mathbf{q}, \mathbf{q}')$. An expansion of the action in powers of $\hbar$ (stationary phase method) leads to a stochastic equation for the particle motion taking into account the backreaction as well as fluctuations caused by the quantum electromagnetic field [142]. We derive this stochastic equation in another way in the next section. We extend the formulas for the particle's density matrix in an environment of a pure state of the quantum electromagnetic field to mixed states of the electromagnetic field.

9.6 Heisenberg Equations of Motion in QED Environment

A relativistic theory of particles interacting with an electromagnetic field is described by the action

$$\int \mathcal{L} = -\frac{1}{4\mu_0} \int dx \, F_{\mu\nu} F^{\mu\nu} + e \int A^\mu dx_\mu + \int \sqrt{dx^\mu dx_\mu}. \tag{9.65}$$

The formula for the density matrix of relativistic particles in an environment of the vacuum electromagnetic fluctuations in a Lorentz invariant formulation takes the form

$$\rho_t(\mathbf{x}, \mathbf{x}') = \int \mathcal{D}q \mathcal{D}q' \exp(i \int_0^t \sqrt{dq_\mu dq^\mu} - i \int_0^t \sqrt{dq'_\mu dq'^\mu})$$
$$\exp\left(-\frac{e^2}{2}i \int_0^t \int_0^t D_{\mu\nu}^{F,\xi}(q(s) - q'(s')) dq^\mu(s) dq'^\nu(s')\right), \tag{9.66}$$

where the average over the electromagnetic field in the path integral formulation has been performed with the photon propagator $D_{\mu\sigma}^{F,\xi}$ of Eq. (9.22). After a continuation to the Euclidean space we obtain from Eq. (9.66) the polymer representation of Sect. 9.3 (now for a particle evolution instead of the field evolution of Sect. 9.3). The non-relativistic limit of Eq. (9.65) in the thermal state has been studied in [142] under the assumption $\mathbf{px} \simeq 0$. We do such computations in another way later in this section. At this stage let us note that in the non-relativistic limit (9.57) of Eq. (9.66) in the radiation gauge at $\mathbf{px} \simeq 0$ we have

$$i D_{jl}^F(t - t', \mathbf{0}) = \mu_0 (2\pi)^{-3} \int d\mathbf{p} (2|\mathbf{p}|)^{-1} \Lambda^{jl} \exp(-i|\mathbf{p}||t - t'|)$$
$$\simeq i \mu_0 \frac{1}{6\pi} \partial_t \delta(t - t').$$

Here, the integral over the spherical angle $(d\theta d\phi \sin\theta)$ is

$$\int d\theta d\phi \sin\theta \Lambda^{jl} = \frac{4\pi}{3} \delta^{jl}.$$

The term quadratic in velocities in the non-relativistic limit of Eq. (9.66) containing the third order derivatives of q contributes to the effective action of the radiation reaction. The effective action leads to the radiation damping term $\frac{\mu_0 e^2}{6\pi} \frac{d^3 x^l}{dt^3}$ as computed in [142].

The relativistic equations for a particle interacting with an electromagnetic field are

$$m \frac{dx^\mu}{ds} = p^\mu,$$

$$\frac{dp^\mu}{ds} = e F^{\mu\nu} p_\nu, \tag{9.67}$$

and (in the SI units)

$$\partial_\mu F^{\mu\nu} = \mu_0 J^\nu, \tag{9.68}$$

where

$$J^\mu(x) = e \int \delta(x - x(s)) dx^\mu(s).$$

$F^{\mu\nu} = \partial^\mu A^\nu - \partial^\nu A^\mu$ and $ds^2 = dx_\mu dx^\mu$. In the Lorentz gauge

$$\partial_\mu A^\mu = 0$$

the field equation can be written as

$$\partial_\mu \partial^\mu A^\nu = \mu_0 J^\nu. \tag{9.69}$$

Instead of the proper time s we may use as an evolution parameter $x_0 = t$. Then, equations of motion for the momentum can be expressed as

$$\frac{dp^j}{dt} = e E^j + e \epsilon^{jrn} \frac{dx^r}{dt} B^n, \tag{9.70}$$

where

$$p^j = m \left(1 - \left(\frac{d\mathbf{x}}{dt}\right)^2\right)^{-\frac{1}{2}} \frac{dx^j}{dt}$$

and (with $A = (\phi, \mathbf{A})$)

$$\mathbf{E} = -\partial_t \mathbf{A} - \nabla\phi, \tag{9.71}$$

$$\mathbf{B} = \nabla \times \mathbf{A}.$$

The vector potential is determined by particle trajectory

$$(\partial_t^2 - \Delta) A^\mu = \mu_0 e \frac{dx^\mu}{dx^0} \delta(x - x(t)). \tag{9.72}$$

The functional integral of Sect. 9.5 leads to an evaluation of the density matrix, the transition probabilities and particle's trajectory from the stationary point of the effective action. Particle's equations of motion can be derived in another way by an elimination of fields from the system of fields and particles (9.70)–(9.72) in the Heisenberg picture.

We use the retarded Green function to solve Eq. (9.72). We work in momentum space

$$\mathbf{A}(\mathbf{x}) = (2\pi)^{-\frac{3}{2}} \int d\mathbf{k} \exp(i\mathbf{k}\mathbf{x}) \mathbf{A}(\mathbf{k}).$$

In the spatially Fourier transformed version the solution of the field equation (9.72) reads

$$A^\mu(t, \mathbf{k}) = A_f^\mu(t, \mathbf{k}) + A_f^\mu(t, \mathbf{k}), \tag{9.73}$$

where the free part A_f satisfies the wave equation

$$\partial^\mu \partial_\mu A_f^\nu = 0 \tag{9.74}$$

and the part of the vector potential produced by the moving particle is

$$A_I^\mu(t, \mathbf{k}) = \mu_0 e (2\pi)^{-\frac{3}{2}} \int_0^t \frac{\sin(k(t-t'))}{k} \exp(-\mathbf{k}\mathbf{x}(t')) \frac{dx^\mu}{dt'} dt'. \tag{9.75}$$

We can rewrite particle equations of motion (9.70) inserting the decomposition (9.73)

$$\begin{aligned}\frac{dp^j}{dt} &= eE_I^j + e\epsilon^{jrn}\frac{dx^r}{dt}B_I^n + eE_f^j + e\epsilon^{jrn}\frac{dx^r}{dt}B_f^n \\ &\equiv eE_I^j + e\epsilon^{jrn}\frac{dx^r}{dt}B_I^n + N_E^j + \epsilon^{jrn}\frac{dx^r}{dt}N_B^n, \end{aligned} \tag{9.76}$$

where the first term (depending on E_I and B_I) modifies equations of motion owing to the interaction of the particle with the electromagnetic field produced by its motion (backreaction). The interaction part E_I and B_I is classical if the effects of non-commutativity of the time dependent coordinates $x(t)$ are neglected. In the semi-classical approximation we assume that $x(t)$ and $\frac{dx}{dt}$ are commuting variables (otherwise the meaning of the products in Eq. (9.70) would have to be explained; the commutators of $x(t)$ would produce terms proportional to $\hbar$ vanishing in the semi-classical approximation). The part coming from the free field solution seems arbitrary from the point of view of Eqs. (9.70) and (9.72). However, in the relativistic quantum field theory of an interaction of fields with particles the field vacuum fluctuations are not negligible. They have some probability distribution. We treat the free part as a noise. In the momentum space

$$N_E^j(t, \mathbf{k}) = eE_f^j(t, \mathbf{k}), \tag{9.77}$$

$$N_B^j(t, \mathbf{k}) = eB_f^j(t, \mathbf{k}) \tag{9.78}$$

The noise is gauge independent. We can study it in an arbitrary gauge. The convenient one is the radiation gauge when $A_f^0 = 0$ and

$$\partial_k A_f^k = 0.$$

(when $A_0 = 0$ this radiation gauge condition agrees with the Lorentz gauge used for a free electromagnetic field (9.74) and for A_I in Eq. (9.72)). We can choose the radiation gauge because A_f^μ satisfies the Maxwell equations (9.74) without source. In this gauge

$$E_f^k = -\partial_t A_f^k. \tag{9.79}$$

Then, the solution of the wave equation (9.74) is

$$A_f^j(t, \mathbf{k}) = A^j(0, \mathbf{k}) \cos(kt) + k^{-1} \sin(kt) \Pi^j(0, \mathbf{k}), \tag{9.80}$$

where $(A^j(0, \mathbf{k}), \Pi^j(0, \mathbf{k}))$ are the initial values of the vector potential and its canonical momentum.

The initial values of the electromagnetic field in Eq. (9.80) are arbitrary. They usually are assumed to be zero. However, such an assumption would not agree with quantum electrodynamics of Sect. 9.5 as treated in Eqs. (9.58)–(9.64). The vacuum fluctuations have observable effects (the Lamb shift [173]). So if we considered in Eq. (9.70) an electron motion in the hydrogen with the potential $\phi = -\frac{e}{r}$ then we would have to include A_f because the electromagnetic field is not produced solely by particle motion but also by vacuum fluctuations (see [188]). In this section we assume that the quantum free electromagnetic field has thermal distribution (as the cosmic microwave background radiation or the electromagnetic field in a cavity in a thermal equilibrium). The quantum free electromagnetic field at finite temperature has the correlation function determined by the density matrix

$$\rho = \Big(Tr(\exp(-\beta H_0))\Big)^{-1} \exp(-\beta H_0).$$

The electromagnetic field correlations can be calculated in the same way as the correlation function for the scalar field obtained in Sect. 4.1

$$
\begin{aligned}
G^{jl}(t, \mathbf{x}; t', \mathbf{x}') &=< A^j(t, \mathbf{x}) A^l(t', \mathbf{x}') >= Tr(\rho A^j(t, \mathbf{x}) A^l(t', \mathbf{x}')) \\
&= \mu_0 \hbar (2\pi)^{-3} \int d\mathbf{k} k^{-1} \cos(\mathbf{k}(\mathbf{x} - \mathbf{x}')) \Lambda^{jl} \\
&\times \Big(\cos(k(t - t')) \coth(\tfrac{1}{2}\beta k \hbar) - i \sin(k(t - t')) \Big),
\end{aligned} \tag{9.81}
$$

where

$$\Lambda^{jl}(k^{-1}\mathbf{k}) = \delta^{jl} - k^{-2} k^j k^l \tag{9.82}$$

comes from the radiation gauge condition. In Eq. (9.81) we write explicitly the Planck constant $\hbar$ in order to study the classical limit.

Note that (as it should be as $\partial_t \mathbf{A} = \mu_0 \Pi$) from Eq. (9.81)

$$[\partial_t A_j(t, \mathbf{x}), A_k(t, \mathbf{y})] = -i \Lambda_{jk} \mu_0 \hbar \delta(\mathbf{x} - \mathbf{y}), \tag{9.83}$$

where on the rhs $\Lambda_{jk} = \delta_{jk} - \partial_j \partial_k \Delta^{-1}$. We are mainly interested of whether the noise is large, but from the commutation relations (9.83) it follows that the electric noise and the magnetic noise cannot be simultaneously arbitrarily small [151, 166].

When $\beta \to \infty$ then we get the quantum electromagnetic field in the radiation gauge at zero temperature in the vacuum state. When $\hbar\beta \to 0$ then

$$< A^j(t, \mathbf{k})A^l(t', \mathbf{k}') >= \mu_0(2\pi)^{-3}\beta^{-1}k^{-2}\Lambda^{jl}\delta(\mathbf{k} + \mathbf{k}')\cos(k(t - t')) \quad (9.84)$$

describes classical field theory in a thermal equilibrium (e.g. electromagnetic field in a cavity). Then, the free modes are distributed according to the classical Gibbs distribution

$$\exp(-\beta H_0) \qquad (9.85)$$

with $\beta^{-1} = k_B T$ and

$$H_0 = \frac{1}{2\mu_0} \int d\mathbf{x}(\mathbf{E}^2 + \mathbf{B}^2). \qquad (9.86)$$

In the classical theory

$$\begin{aligned} &< N_E^j(t, \mathbf{k})N_E^l(t', \mathbf{k}') >=< N_B^j(t, \mathbf{k})N_B^l(t', \mathbf{k}') > \\ &= \mu_0 e^2 \beta^{-1}\Lambda^{jl}\delta(\mathbf{k} + \mathbf{k}')\cos(k(t - t')), \end{aligned} \qquad (9.87)$$

$$< N_E^j(t, \mathbf{k})N_B^l(t', \mathbf{k}') >= \mu_0 k_m k^{-1}\epsilon^{lmj} e^2 \beta^{-1}\delta(\mathbf{k} + \mathbf{k}')\sin(k(t - t')). \quad (9.88)$$

In coordinate space on the particle trajectory

$$\begin{aligned} &< N_E^j(t, \mathbf{x}(t))N_E^l(t', \mathbf{x}(t')) > \\ &= e^2\beta^{-1}\mu_0(2\pi)^{-3} \int d\mathbf{k}\Lambda^{jl}\exp(i\mathbf{k}(\mathbf{x}(t) - \mathbf{x}(t')))\cos(k(t - t')) \end{aligned} \qquad (9.89)$$

which can also be expressed as

$$\begin{aligned} &< N_E^j(t, \mathbf{x}(t))N_E^l(t', \mathbf{x}(t')) >= \mu_0(2\pi)^{-3}e^2 4\pi\beta^{-1} \\ &\int_0^\infty dk k^2 \Lambda^{jl}(k^{-1}\nabla)\sin(k|\mathbf{x}(t) - \mathbf{x}(t')|)(k|\mathbf{x}(t) - \mathbf{x}(t')|)^{-1}\cos(k(t - t')), \end{aligned} \qquad (9.90)$$

where

$$\begin{aligned} \Lambda^{jl}(k^{-1}\nabla)\sin(k|\mathbf{x}|)(k|\mathbf{x}|)^{-1} = &\left((k|\mathbf{x}|)^{-2}\cos(k|\mathbf{x}|) - (k|\mathbf{x}|)^{-3}\sin(k|\mathbf{x}|)\right)\delta^{jl} + \\ &|\mathbf{x}|^{-2}x^j x^l\left(3(k|\mathbf{x}|)^{-3}\sin(k|\mathbf{x}|) - (k|\mathbf{x}|)^{-1}\sin(k|\mathbf{x}|) - 3(k|\mathbf{x}|)^{-2}\cos(k|\mathbf{x}|)\right). \end{aligned} \qquad (9.91)$$

We can compute the integral (9.90) using the formula

$$\int\limits_0^\infty dk k^{-1} \sin(ku) = \frac{\pi}{2}\epsilon(u) \qquad (9.92)$$

and its derivatives over u. We note (as expected) that the noise is concentrated on the light cone.

If $\mathbf{x}(t) = \mathbf{x}(t') = 0$ (i.e., for small $k\mathbf{x}$) then

$$< N_E^j(t, \mathbf{0}) N_E^l(t', \mathbf{0}) > = \mu_0 (2\pi)^{-3} e^2 \tfrac{4}{3} \delta^{jl} \pi^2 \beta^{-1} \partial_t \partial_{t'} \delta(t - t'). \qquad (9.93)$$

In computation of the integral (9.93) we used the average over the spherical angle

$$\int d\phi d\theta \sin\theta \Lambda^{jl} = \frac{4\pi}{3}\delta^{jl}.$$

We return now to Eq. (9.76). This is a non-local and non-linear integro-differential equation. It simplifies in the non-relativistic limit and in the limit of small $\mathbf{x}$ (small in comparison to $\frac{1}{k}$). In order to derive this simplification let us calculate from Eqs. (9.71) and (9.75)

$$e E_I^j(t) = \mu_0 e^2 (2\pi)^{-3} \int d\mathbf{k} \Lambda^{jl} \int\limits_0^t \cos(k(t - t')) \frac{dx^l}{dt'} dt', \qquad (9.94)$$

where we have neglected $\mathbf{x}(t)$ in the exponential of Eq. (9.75) as small for $|\mathbf{x}| << \frac{1}{k}$. The term $\nabla\phi$ from Eq. (9.71) when integrated over $d\mathbf{k}$ at $x(t) = 0$ in Eq. (9.75) gives 0. In the non-relativistic limit the A^0 component becomes time independent. Hence, the Lorentz gauge condition applied to derive Eq. (9.72), becomes the Coulomb condition $\partial^k A_k = 0$. For this reason there is the projection operator Λ in Eq. (9.94) ensuring $k_j E_I^j = 0$. The integration in Eq. (9.94) using $d\mathbf{k} = dk k^2 d\theta d\phi \sin\theta$ and $k^2 \cos(k(t - t')) \to -\partial_t^2 \cos(k(t - t'))$ leads to an integral producing the δ-function as

$$\int\limits_0^\infty dk \cos(k(t - t')) = \pi\delta(t - t'). \qquad (9.95)$$

As a result in Eq. (9.94) we obtain

$$\frac{1}{4\pi}\frac{2}{3}\mu_0 e^2 \frac{d^3 x^j}{dt^3} \qquad (9.96)$$

The magnetic part in Eq. (9.76) is negligible in the non-relativistic limit so that Eq. (9.76) takes the form

$$m\frac{d^2x_j}{dt^2} = \frac{1}{4\pi}\frac{2}{3}\mu_0 e^2\frac{d^3x_j}{dt^3} + \gamma\frac{d^2w_j^e}{dt^2},$$ (9.97)

where $\mathbf{w}^e(t)$ is the Brownian motion as follows from the correlation function (9.93) $(\gamma\frac{d^2w_j^e}{dt^2} = N_e^j)$. We add the index e to $\mathbf{w}$ in order to distinguish it from the Brownian motion applied in the Feynman integral in Eqs. (5.3)–(5.4)). The noise correlation function (9.93) has been obtained in the limit when the electromagnetic field is described by a classical Gibbs factor (9.85)–(9.86) with the non-relativistic long wave approximation which gives $\gamma^2 = \mu_0\frac{1}{6\pi}e^2\beta^{-1}$. The result (9.97) leads to a fluctuation-dissipation relation discussed in [142] which can be applied to the theory of spectral lines at the temperature β^{-1}. The dissipation must be accompanied by noise if the equilibrium is to be preserved. The radiation damping term in classical electrodynamics is discussed in [162, 207].

We can obtain the non-relativistic version of the equation of motion (9.76) of a particle in a quantum state of the thermal electromagnetic field if we calculate the quantum noise from Eq. (9.81). In quantum theory the average of the quantum noise (9.77)–(9.80) in the thermal state is

$$\begin{aligned}< N_E^j(t,\mathbf{k})N_E^l(t',\mathbf{k}') > &=< N_B^j(t,\mathbf{k})N_B^l(t',\mathbf{k}') >\\ &= \mu_0 e^2(2\pi)^{-3}\hbar\Lambda^{jl}\delta(\mathbf{k}+\mathbf{k}')k\\ &\times\Big(\coth(\tfrac{1}{2}\beta\hbar k)\cos(k(t-t')) - i\sin(k(t-t'))\Big),\end{aligned}$$ (9.98)

$$\begin{aligned}< N_E^j(t,\mathbf{k})N_B^l(t',\mathbf{k}') > &= \mu_0 c^{-2}\hbar k_m k^{-1}\epsilon^{lmj}e^2\delta(\mathbf{k}+\mathbf{k}')\\ &\times\Big(-\coth(\tfrac{1}{2}\beta\hbar k)\sin(k(t-t')) - i\cos(k(t-t'))\Big).\end{aligned}$$ (9.99)

In the computation of spatial correlations of the noise at the temperature β^{-1} we use the formula [115]

$$\int_0^\infty du\,\sin(au)\Big(\exp(\beta u) - 1\Big)^{-1} = \frac{\pi}{2\beta}\coth\Big(\frac{\pi a}{\beta}\Big) - \frac{1}{2a}.$$ (9.100)

Now

$$\begin{aligned}&\tfrac{1}{2}(< N_E^j(t,\mathbf{x}(t))N_E^l(t',\mathbf{x}(t')) > + < N_E^l(t',\mathbf{x}(t'))N_E^j(t,\mathbf{x}(t)) >)\\ &= \mu_0 e^2(2\pi)^{-3}\pi\int_0^\infty dk k^2\Lambda^{jl}\sin(k|\mathbf{x}(t) - \mathbf{x}(t')|)(k|\mathbf{x}(t) - \mathbf{x}(t')|)^{-1}\\ &\cos(k(t-t'))\hbar k\Big(1 + 2(\exp(\beta\hbar k) - 1)^{-1}\Big)\\ &\equiv (\mathcal{D} + \mathcal{D}_T)(t,\mathbf{x}(t); t',\mathbf{x}(t')).\end{aligned}$$ (9.101)

Here

$$\Lambda^{jl} = \Lambda^{jl}(k^{-1}\nabla_\mathbf{x}).$$

The first term on the rhs of Eq. (9.101) describes highly singular quantum vacuum fluctuations of the electric field proportional to $\hbar$ which vanish at large space-time separations. The second term is

$$
\begin{aligned}
& \mathcal{D}_T^{jl}(t, \mathbf{x}(t)); t', \mathbf{x}(t')) \\
&= \mu_0 \hbar e^2 \frac{\pi}{2} \int_0^\infty dk k^2 \Lambda^{jl}(k^{-1}\nabla) \sin(k|\mathbf{x}(t) - \mathbf{x}(t')|)|\mathbf{x}(t) - \mathbf{x}(t')|^{-1} \\
& \times \cos(k(t - t'))(\exp(\beta \hbar k) - 1)^{-1} = \mu_0 \hbar e^2 \delta^{jl} \frac{\pi}{4} |\mathbf{x}(t) - \mathbf{x}(t')|^{-1} \partial_t \partial_{t'} \\
& \times \left(\frac{\pi}{\beta\hbar} \coth\left(\frac{\pi}{\beta\hbar}(t - t' + |\mathbf{x}(t) - \mathbf{x}(t')|) \right) \right. \\
& \quad - \frac{\pi}{\beta\hbar} \coth\left(\frac{\pi}{\beta\hbar}(t - t' - |\mathbf{x}(t) - \mathbf{x}(t')|) \right) \\
& \quad \left. + \left(t - t' - |\mathbf{x}(t) - \mathbf{x}(t')| \right)^{-1} - \left(t - t' + |\mathbf{x}(t) - \mathbf{x}(t')| \right)^{-1} \right) + \cdots
\end{aligned}
\tag{9.102}
$$

In Eq. (9.102) we write only the least singular term corresponding to the first term (proportional to δ^{jl}) in Eq. (9.91). The time derivatives in Eq. (9.102) do not apply to $\mathbf{x}(t)$. Usually the spectrum of noise of Eq. (9.98) is compared with experiments [49] [152]. However, the space-time correlations (9.101)–(9.102) may supply an additional information. The correlations of noise defined in Eq. (9.80) can be calculated in any state providing an information about the state of the electromagnetic field.

Returning to Eq. (9.76) we note that this is an operator equation. This means that $x(t)$ must be treated as an operator because on the rhs we have an operator noise. For small x both sides in Eq. (9.76) are linear in x. Then, the problem of an interpretation of a product of operators does not appear. However, if we consider the space-time dependence of the noise in Eq. (9.76) then discussing the space-time correlations of $x(t)$ we should treat the non-commutativity of $x(t)$ in the noise-correlations (9.101)–(9.102) with more care (calculating these correlations we assumed that $x(t)$ and $x(t')$ commute).

9.7 Noise in the Squeezed State

We are interested whether the quantum nature of the electromagnetic field can be confirmed in experiments involving an interaction of charged particles with the electromagnetic field. The confirmation comes from the photo-detection [181] or from the Lamb shift [173] on the basis of the theory of Sect. 9.5. It would be rather difficult to follow an electron trajectory in order to check the Heisenberg equations of motion of Sect. 9.6. If the interaction is weak then instead of the variation of particle trajectory we could observe the photon noise. This is the idea suggested in ref. [194] in the case of gravitons. In this section we consider a similar question for photons. The squeezed states of light have been produced in quantum optics [101, 231]. Using the definition of the noise (9.77) we can compute the correlation functions of the noise in any state. We calculate here the correlation functions of the noise in the Gaussian state

$$\psi_0^\Gamma = N \exp\left(-\frac{1}{2\hbar}\Lambda\mathbf{A}\Gamma_0\Lambda\mathbf{A}\right), \tag{9.103}$$

where N is a normalization factor. We assume that in a squeezed state $\Gamma_0 + \Gamma_0^* \to 0$. We have

$$
\begin{aligned}
(\psi_0^\Gamma, A_t^j(\mathbf{k})A_{t'}^l(\mathbf{k}')\psi_0^\Gamma) &= (\psi_0^\Gamma, (A^j(\mathbf{k})\cos(kt) + k^{-1}\sin(kt)\Pi^j(\mathbf{k})) \\
&\times (A^l(\mathbf{k}')\cos(k't') + k'^{-1}\sin(k't')\Pi^l(\mathbf{k}'))\psi_0^\Gamma) \\
&= (\psi_0^\Gamma, A^j(\mathbf{k})\cos(kt) + k^{-1}\sin(kt)i\Gamma(\mathbf{k})A^j(-\mathbf{k})) \\
&\times (A^l(\mathbf{k}')\cos(k't') + k'^{-1}\sin(k't')i\Gamma(\mathbf{k}')A^l(-\mathbf{k}'))\psi_0^\Gamma) \\
&- i\hbar k^{-1}\Lambda^{jl}\sin(kt)\cos(k't')\delta(\mathbf{k} - \mathbf{k}') \\
&+ \hbar k^{-1}\Lambda^{jl}\sin(kt)k^{-1}\sin(k't')\Gamma(k)\delta(\mathbf{k} + \mathbf{k}').
\end{aligned}
\tag{9.104}
$$

Now

$$(\psi_0^\Gamma)^*\psi_0^\Gamma = \exp\left(-\frac{1}{2\hbar}\Lambda\mathbf{A}(\Gamma_0 + \Gamma_0^*)\Lambda\mathbf{A}\right). \tag{9.105}$$

Hence,

$$(\psi_0^\Gamma, A^j(\mathbf{k})A^l(\mathbf{k}')\psi_0^\Gamma) = \hbar\Lambda^{jl}(\Gamma + \Gamma^*)^{-1}\delta(\mathbf{k} + \mathbf{k}'). \tag{9.106}$$

For a small $\Gamma + \Gamma^*$ from Eqs. (9.104) and (9.106)

$$(\psi_0^\Gamma, A_t^j(\mathbf{k})A_{t'}^l(\mathbf{k}')\psi_0^\Gamma) \simeq \hbar\Lambda^{jl}\cos(kt)\cos(k't')(\Gamma_0 + \Gamma_0^*)^{-1}\delta(\mathbf{k} + \mathbf{k}') \tag{9.107}$$

because the large $(\Gamma_0 + \Gamma_0^*)^{-1}$ is dominating . So the electric noise in the state (9.103) for a small $\Gamma_0 + \Gamma_0^*$ is

$$(\psi_0^\Gamma, N_E^j(t, \mathbf{k})N_E^l(t', \mathbf{k}')\psi_0^\Gamma) = e^2\hbar\Lambda^{jl}\cos(kt)\cos(k't')(\Gamma_0 + \Gamma_0^*)^{-1}k^2\delta(\mathbf{k} + \mathbf{k}'). \tag{9.108}$$

The n-point correlation functions in the state (9.103) are products of the two-point functions of the noise (in the limit of strong squeezing). If the n-point functions do not have this property then the measured state is not Gaussian. Measuring the correlation functions we could determine the state of the quantum electromagnetic field.

9.8   Feynman Formula in QED with an Axion

The field of an axion appeared as a solution of the CP violation problem in quantum chromodynamics [64, 244]. The idea is to explain the weak CP violation as a result of the coupling $\phi\epsilon_{\mu\nu\sigma\rho}F^{\mu\nu}F^{\sigma\rho}$ of the gauge field to a scalar field ϕ (axion). The experimental weakness of the CP violation can be viewed as a consequence of the dynamics. The axion particles have been proposed as candidates for the dark matter

[200]. The addition of the axion field to the Hamiltonian leads to an interesting model of gauge theory. It is like an addition of a gauge potential to the Hamiltonian defined in an infinite number of dimensions.

We consider the Lagrangian

$$\mathcal{L}(A, \phi) = -\frac{1}{4\mu_0} F_{\mu\nu} F^{\mu\nu} + \phi \epsilon_{\mu\nu\sigma\rho} F^{\mu\nu} F^{\sigma\rho} + \mathcal{L}(\phi), \qquad (9.109)$$

where $\mathcal{L}(\phi)$ is the Lagrangian for the ϕ-field. We restrict ourselves to Abelian gauge theory. We work in the temporal gauge $A^0 = 0$. Then, the Lagrangian (9.109) can be written as

$$\mathcal{L}(A, \phi) = \frac{1}{2\mu_0}((\partial_t \mathbf{A})^2 - (\nabla \times \mathbf{A})^2) + \phi \epsilon_{jkl} \partial_t A^j F^{kl} + \mathcal{L}(\phi). \qquad (9.110)$$

The solution of the CP-problem requires the choice of a periodic $\mathcal{L}(\phi)$ (i.e., there exists f such that $\mathcal{L}(\phi) = \mathcal{L}(\phi + f)$). The canonical momentum is

$$\Pi_j^F = \frac{1}{\mu_0} \partial_t A_j + \phi \epsilon_{jkl} F^{kl}. \qquad (9.111)$$

Hence, the Hamiltonian reads

$$H = \frac{\mu_0}{2} \sum_j (\Pi_j^F - \phi \epsilon_{jkl} F^{kl})^2 + \frac{1}{2\mu_0}(\nabla \times \mathbf{A})^2 + H(\phi) \equiv H(A, \phi) + H(\phi). \qquad (9.112)$$

The Hamiltonian in Eq. (9.112) (with $H(\phi) = 0$ and $\Pi_j^F = -i\frac{\delta}{\delta \Pi^F}$) is analogous to the Hamiltonian for a particle in a gauge potential in the sense that if $H(\mathbf{A}, 0)\psi = E\psi$ then for the state

$$\psi' = \exp\left(i \int d\mathbf{x} \phi \epsilon^{jkl} F_{kl} A_j\right) \psi \qquad (9.113)$$

we have

$$H(\mathbf{A}, \phi)\psi' = E\psi'$$

In this sense the interaction with ϕ does not change energy.

Neglecting for the moment the $H(\phi)$ part we can write the Feynman formula for the Hamiltonian evolution

$$\psi_t(\mathbf{A}) = E\left[\exp\left(\frac{i}{\hbar} \int_0^t \phi(t - s, \mathbf{x}) \epsilon^{jkl} F_{kl}(s, \mathbf{x}) dA^j(s, \mathbf{x}) dx\right) \psi(\mathbf{A}_t(\mathbf{A})) \right], \qquad (9.114)$$

where

$$dA_t^j = -i\omega A_t^j dt + \sqrt{i\hbar\mu_0} \Lambda^{jk} dW_t^k. \qquad (9.115)$$

In order to see the effect of an axion on the correlation functions we would have to calculate the functional integral

$$\left\langle (\psi_t, A^r(\mathbf{x})A^n(\mathbf{x}')\psi_t) \right\rangle, \tag{9.116}$$

where $< .. >$ means an average over the axion field ϕ. Calculation of such an average would be useful for an experimental verification of the axion hypothesis.

9.9 Decoherence in an Environment of Photons

We discuss again a particle interacting with quantum electromagnetic field as in Sect. 9.5. We consider an initial state $\psi(\mathbf{x}, A) = \chi(\mathbf{x})\psi_0^g(\mathbf{x})\chi_{em}(A)\psi_{em}^g(A)$, where $\psi_0^g(\mathbf{x})$ is the particle ground state and $\psi_{em}^g(\mathbf{A})$ the electromagnetic ground state (we added an index em to wave functions of the electromagnetic field). The unitary time evolution takes the form

$$\psi_t(\mathbf{x}, \mathbf{A}) = \chi(\mathbf{x}_t(\mathbf{x}, A))\chi_{em}(\mathbf{A}_t(\mathbf{A}, \mathbf{x}))\psi_0^g(\mathbf{x})\psi^g(\mathbf{A}), \tag{9.117}$$

where $\mathbf{x}_t(\mathbf{x}, \mathbf{A})$ and $\mathbf{A}_t(\mathbf{A}, \mathbf{x})$ are the solutions of the Heisenberg equations of motion (9.75)–(9.76). We take an average $< ... >$ over the states of the electromagnetic field defining the density matrix

$$\rho_t(\mathbf{x}, \mathbf{x}') = < \psi_t^*(\mathbf{x}, \mathbf{A})\psi_t(\mathbf{x}', \mathbf{A}) > . \tag{9.118}$$

If the averaging in Eq. (9.118) is with respect to a thermal state of the electromagnetic field then according to Sect. 9.6 the Heisenberg equation of motion can be treated as a stochastic equation for a particle with the Brownian noise $\frac{d\mathbf{w}^e}{dt}$. Let us first consider the simplest case of a free particle interacting with photons. Then, in Eq. (9.97) in the first approximation we can neglect the backreaction term as $\frac{d^3\mathbf{x}}{dt^3} = 0$ for a free particle. Using the path integral in the formulation of Sect. 5.3 and 5.4 we can express the solution of the Schrödinger equation for a quantum particle moving in the noise of Eq. (9.97) as

$$\chi_t(\mathbf{x}, \mathbf{w}^e) = E_w[\chi(\mathbf{q}_t(\mathbf{x}))], \tag{9.119}$$

where

$$\mathbf{q}_t(x) = \mathbf{x} + \sigma\mathbf{w}_t + \frac{\gamma}{m}\mathbf{w}_t^e. \tag{9.120}$$

Here $\sigma = \sqrt{\frac{i\hbar}{m}}$, $\mathbf{w}$ is the Brownian motion representing the Schrödinger dynamics of Sect. 5.3 and the expectation value in Eq. (9.119) is with respect to $\mathbf{w}$. The Brownian motion $\mathbf{w}^e$ is discussed in Sect. 9.5 as a result of photon scattering ($\mathbf{w}^e$ and $\mathbf{w}$ are independent Brownian motions). For a free particle $\psi^g(\mathbf{x}) = 1$ (the ground state is not normalizable). The equation for the density matrix can be obtained by a differentiation of Eq. (9.118) where the average over the electromagnetic field is expressed by an average over the noise $\mathbf{w}^e$ (according to Sect. 9.6). We have ($E_e[...]$ denotes an expectation value with respect to $\mathbf{w}^e$)

$$
d\rho_t(\mathbf{x}, \mathbf{x}') = E_e\Big[E_w[d\chi^*(\mathbf{q}_t(\mathbf{x}))]E_w[\chi(\mathbf{q}_t(\mathbf{x}'))]\Big]
$$
$$
+E_e\Big[E_w[\chi^*(\mathbf{q}_t(\mathbf{x}))]E_w[d\chi(\mathbf{q}_t(\mathbf{x}'))]\Big] + E_e\Big[E_w[d\chi^*(\mathbf{q}_t(\mathbf{x}))]E_w[d\chi(\mathbf{q}_t(\mathbf{x}'))]\Big].
$$
$$
(9.121)
$$

The Ito formula (5.67) from Sect. 5.4 gives

$$
d\chi(\mathbf{q}_t) = \nabla\chi d\mathbf{q}_t + \frac{1}{2}(d\mathbf{q}_t\nabla)^2\chi. \tag{9.122}
$$

Then, in order to calculate the rhs of Eq. (9.121) it is sufficient to multiply $d\chi$ of Eq. (9.122) according to the rules $dw^j dw^k = \delta^{jk}dt, dw_j dw_k^e = 0$ and $dw_j^e dw_k^e = \delta_{jk}dt$.
Equation (9.121) can be expressed as

$$
\partial_t\rho = \frac{i}{\hbar}[H_0, \rho] + \frac{1}{2\hbar^2 m^2}\gamma^2(\mathbf{L}^+\mathbf{L}\rho + \rho\mathbf{L}^+\mathbf{L} - 2\mathbf{L}\rho\mathbf{L}^+), \tag{9.123}
$$

where $\mathbf{L} = -i\hbar\nabla$ and

$$
H_0 = -\frac{\hbar^2}{2m}\nabla^2.
$$

When $\mathbf{L}$ is a Hermitian operator then Eq. (9.123) can also be written in the form

$$
\partial_t\rho = \frac{i}{\hbar}[H_0, \rho] - \frac{\gamma^2}{2m^2\hbar^2}[\mathbf{L}, [\mathbf{L}, \rho]]. \tag{9.124}
$$

Equation (9.123) represents a general form of the dissipative dynamics (Lindblad equation [19, 114, 176]) which preserves the positivity and the trace of the density matrix. In the momentum basis Eq. (9.124) can easily be solved with the result

$$
< \mathbf{p}|\rho_t|\mathbf{p}' >= \exp\Big(\frac{i}{2m\hbar}t(\mathbf{p}^2 - \mathbf{p}'^2) - \frac{1}{2}t\gamma^2 m^{-2}\hbar^{-2}(\mathbf{p} - \mathbf{p}')^2\Big)\rho_0(\mathbf{p}). \tag{9.125}
$$

The solution (9.125) shows that the density matrix becomes diagonal in momenta for a large time (decoherence, [249]). We can extend Eq. (9.120) to general particle dynamics in a state ψ_t influenced by the environmental noise coming from the scattering of thermal photons. This equation reads

$$
d\mathbf{q}_s = \frac{i\hbar}{m}\nabla \ln \psi_{t-s}^\Gamma ds + \sigma d\mathbf{w} + \gamma d\mathbf{w}_s^e. \tag{9.126}
$$

Repeating the calculation (9.121) we obtain the Lindblad equation (9.123).

9.10   Entropy of Gaussian Wigner States

The entropy S of a quantum mixed state ρ has been defined by von Neumann as $S = -Tr(\rho \ln \rho)$. An investigation of the entropy became an interesting research problem in relation to decoherence in quantum mechanics (optimal uncertainty [34, 62], quantum computations [174]) and in cosmology (thermodynamics of the Universe [40, 41]). If $\rho = \sum p_j |j><j|$, where p_j is a probability of the state $|j>$ then $S = -\sum p_j \ln p_j$. This is the expression known from classical statistical mechanics [113] and information theory (Shannon) [174]. In this section we discuss the entropy of states of a scalar field but similar formulas can be obtained for fields with more components as vector and tensor fields which are considered as an environment of other systems in this chapter and in Chap. 11. The Gaussian Wigner function defined in Sect. 8.7 has the property of a probability density function in the phase space of a classical system. For such a probability distribution we can define the entropy as Boltzmann's H-function [113]

$$S = - \int d\phi d\Pi p(\phi, \Pi) \ln p(\phi, \Pi). \tag{9.127}$$

If we have a classical Hamiltonian system with a Hamiltonian $H(\phi, \Pi)$

$$\frac{d\phi}{dt} = \Pi, \tag{9.128}$$

$$\frac{d\Pi}{dt} = -\frac{\delta H}{\delta \phi}, \tag{9.129}$$

then by a formal integration by parts we can show that $\frac{dS}{dt} = 0$. Hence, as expected for a deterministic classical system, the entropy does not change. As we discussed in Sect. 9.9 quantum systems in a heat bath (environment) of other particles can be treated as a stochastic classical system. Let us consider a free particle in the environment of thermal photons described by Eq. (9.120). In the limit $\hbar \to 0$ the particle distribution p is given by the solution of the diffusion equation

$$\partial_t p = \frac{\gamma^2}{2m^2} \Delta p. \tag{9.130}$$

By differentiation of $S = - \int d\mathbf{x} p(\mathbf{x}) \ln p(\mathbf{x})$ we can show that the particle entropy in an environment of photons is increasing [206]

$$\partial_t S = \frac{\gamma^2}{2m^2} \int d\mathbf{x} p \nabla \ln p \nabla \ln p. \tag{9.131}$$

The states of quantum fields in an environment of noise (in the limit $\hbar \to 0$) can also be described as states of classical stochastic systems by an addition of the Wiener noise W to the rhs of Eq. (9.129) (see, e.g., [141])

$$\frac{d\Pi}{dt} = -\frac{\delta H}{\delta \phi} + \nu \frac{dW}{dt},$$

where ν is the diffusion constant and $W(t, \mathbf{x})$ is the Wiener process discussed in Sect. 8.5. The probability distribution $p(\phi, \Pi)$ satisfies the Fokker-Planck equation

$$\partial_t p = \int d\mathbf{x} \left(\Pi(\mathbf{x}) \frac{\delta p}{\delta \phi(\mathbf{x})} - \frac{\delta H}{\delta \phi(\mathbf{x})} \frac{\delta p}{\delta \Pi(\mathbf{x})} + \frac{\nu^2}{2} \frac{\delta^2 p}{\delta \Pi(\mathbf{x}) \delta \Pi(\mathbf{x})} \right). \qquad (9.132)$$

By direct calculations we obtain an analog of Eq. (9.131) in an infinite number of dimensions

$$\partial_t S = \frac{\gamma^2}{2} \int d\phi d\Pi p(\phi, \Pi) \int d\mathbf{x} \frac{\delta \ln p}{\delta \Pi(\mathbf{x})} \frac{\delta \ln p}{\delta \Pi(\mathbf{x})}. \qquad (9.133)$$

If p is Gaussian then the integrals (9.127) as well as (9.133) are well-defined as Gaussian integrals in an infinite number of dimensions in the sense of Sect. 1.3.

We consider now the entropy of the Gaussian Wigner function (8.61). For Gaussian $p(\phi, \Pi)$ the functional integral (9.127) is again Gaussian.

$$S = \frac{1}{2} \int dX \exp\left(-\frac{1}{2} X^T G X \right) X^T G X \det G^{\frac{1}{2}} - \frac{1}{2} \ln \det G \simeq -\frac{1}{2} \ln \det G$$
$$(9.134)$$

In this formula $X^T = (\phi, \Pi)$, G is an operator which has matrix elements G_{ij}, where $i, j = 1, 2$ and G_{ij} are operators with the integral kernels (Sect. 3.4) $G_{ij}(\mathbf{x} - \mathbf{y})$. The first term on the rhs of Eq. (9.134) is just a constant which does not depend on the state under consideration (depending on G). We investigate the second term $-\frac{1}{2} \ln \det G$. If we have an analytic function $f(G) = \sum_n c_n G^n$ of an operator G with the integral kernel $G(\mathbf{x} - \mathbf{y})$ then

$$\begin{aligned} Tr f(G) &= Tr\left(\sum_n c_n G^n \right) = (2\pi)^{-3} \int d\mathbf{x} \int d\mathbf{k} \sum_n c_n G(\mathbf{k})^n \\ &= (2\pi)^{-3} \int d\mathbf{x} \int d\mathbf{k} f(G(\mathbf{k})), \end{aligned} \qquad (9.135)$$

where

$$G(\mathbf{x} - \mathbf{y}) = (2\pi)^{-3} \int d\mathbf{k} \exp(i\mathbf{k}(\mathbf{x} - \mathbf{y})) G(\mathbf{k}). \qquad (9.136)$$

We check Eq. (9.135) term by term

$$Tr G = \int d\mathbf{x} G(\mathbf{x} - \mathbf{x}) = \int d\mathbf{x} G(0) = \int d\mathbf{x}(2\pi)^{-3} \int d\mathbf{k} G(\mathbf{k}),$$

$$\begin{aligned} Tr G^2 &= \int d\mathbf{x} d\mathbf{y} G(\mathbf{x} - \mathbf{y}) G(\mathbf{y} - \mathbf{x}) \\ &= \int d\mathbf{x}(2\pi)^{-6} \int d\mathbf{y} d\mathbf{k} d\mathbf{q} G(\mathbf{k}) G(\mathbf{q}) \exp(i\mathbf{k}(\mathbf{x} - \mathbf{y}) - i\mathbf{q}(\mathbf{x} - \mathbf{y})) \\ &= \int d\mathbf{x}(2\pi)^{-3} \int d\mathbf{k} G(\mathbf{k})^2 \end{aligned}$$

and similarly for the n-th order term. Hence, using the formula $\ln \det G = \ln \exp(Tr \ln G) = Tr \ln G$ we obtain that the entropy density in Eq. (9.134) is

$$s = S\left(\int d\mathbf{x} \right)^{-1} = -\frac{1}{2}(2\pi)^{-3} \int d\mathbf{k} \ln \det G(\mathbf{k}), \qquad (9.137)$$

where G is the matrix of operators of Eq. (8.61). So, finally the formula for the entropy density of a Wigner function is

$$s = -\frac{1}{2}(2\pi)^{-3} \int d\mathbf{k} \ln \left(G_{11}(\mathbf{k})G_{22}(\mathbf{k}) - G_{12}^2(\mathbf{k}) \right) \qquad (9.138)$$

For a ground state of Eq. (4.40) $G_{12} = 0$, $G_{11} = (\mathbf{k}^2 + m^2)^{\frac{1}{2}}$, $G_{22} = (\mathbf{k}^2 + m^2)^{-\frac{1}{2}}$. The entropy is an infinite constant. For the Gaussian state (8.52) as shown in Sect. 8.7 $\det G = G_{11}G_{22} - G_{12}^2 = \det 4 = const$. Hence, for pure states satisfying the free Schrödinger equation the entropy is a trivial constant.

Summarizing , for pure states satisfying the Schrödinger equation the entropy is a constant. If the quantum system is in an environment of a heat bath the entropy can increase (see the calculations for an inflationary Universe in [40, 41]). The entropy is increasing for stochastic systems of Eq. (9.133). For quantum dissipative systems described by the Lindblad equation the entropy is also increasing [227].

9.11 Exercises

9.1 Show that if $\partial_\mu F^{\mu\nu} = 0$ then by means of the gauge transformation we can choose the radiation gauge $A^0 = 0$ and $\partial_k A^k = 0$.

9.2 Solve Eq. (9.7) by means of the Fourier transform to obtain Eq. (9.8).

9.3 Derive the Feynman rules for the Higgs interaction (9.21).

9.4 Show that one can choose the coordinate gauge (9.40) and prove Eq. (9.41)

9.5 Consider the harmonic oscillator in a quantum electromagnetic field. Calculate in a perturbation expansion (in an electric charge) the transition probability from the ground state to the first excited state of the oscillator using Eq. (9.63) (see [188], Appendix E or [181], Sect. 14)

9.6 Solve Eq. (9.97) perturbatively. Calculate the correlation functions $< x(t)x(t') >$ in the electromagnetic vacuum and in the squeezed state of Sect. 9.8.

9.7 Using the formula (9.52) calculate the one loop electromagnetic effective action resulting from an interaction of the complex scalar field with an electromagnetic field. Use an analytic continuation in time in order to obtain the Euler-Heisenberg result in the Minkowski space-time.

9.8 Using Eq. (9.126) derive the Lindblad equation for a quantum oscillator in its ground state under the influence of photon noise.

Chapter 10
Particle Interaction with Gravitons

Abstract The quantization of gravity is still an unsolved problem. However, thus far in all quantized models there is a perturbative regime when the fields behave as free fields. Such a quantization scheme is based on the observation that there are particles associated to the quantized fields. The detection of gravitational waves gives a support for the conjecture that there is a regime of quantum gravity when the gravitational waves can be treated as a stream of particles. After a short introduction to classical gravity we quantize the linearized gravity in the transverse-traceless gauge. We discuss the geodesic equation in a falling frame of another particle (the geodesic deviation equation) as an analog of the particle equation of motion in quantum electrodynamics. We derive a stochastic equation for a particle moving in an environment of thermal gravitons.

The quantization of gravity is still an unsolved problem. However, thus far in all quantized models there is a perturbative regime when the fields behave as free fields. Such a quantization scheme is based on the observation that there are particles associated to the quantized fields. The detection of gravitational waves [1] gives a support for the conjecture that there is a regime of quantum gravity when the gravitational waves can be treated as a stream of particles. After a short introduction to classical gravity in Sect. 10.1 we quantize in Sect. 10.2 the linearized gravity in the transverse-traceless gauge which is an analog of the radiation gauge of QED. We discuss the geodesic equation in a falling frame of another particle (the geodesic deviation equation) as an analog of the particle equation of motion in QED. Both equations are expressed by the field strength (the curvature tensor in gravity). In a close analogy with QED we derive the stochastic equation for a particle moving in an environment of thermal gravitons (Sects. 10.2 and 10.3). We obtain a formula for the correlation functions of the graviton noise (Sect. 10.4) which could be measured in gravity wave experiments.

© The Author(s), under exclusive license to Springer Nature Switzerland AG 2023   179
Z. Haba, *Lectures on Quantum Field Theory and Functional Integration*,
https://doi.org/10.1007/978-3-031-30712-6_10

10.1   Classical Gravity

The classical Riemannian geometry is defined [213] by the metric tensor

$$ds^2 = g_{\mu\nu}dx^\mu dx^\nu$$

which determines the parallel transport defined by the Christoffel symbol

$$\Gamma^\lambda_{\mu\nu} = \frac{1}{2}g^{\lambda\alpha}(\partial_\mu g_{\alpha\nu} + \partial_\nu g_{\alpha\mu} - \partial_\alpha g_{\mu\nu}).$$

From the Christoffel symbols we can construct the curvature tensor $R^\mu_{\nu\sigma\rho}$

$$R^\mu_{\nu\sigma\rho} = \partial_\sigma \Gamma^\mu_{\nu\rho} - \partial_\rho \Gamma^\mu_{\nu\sigma} + \Gamma^\alpha_{\nu\rho}\Gamma^\mu_{\alpha\sigma} - \Gamma^\alpha_{\nu\sigma}\Gamma^\mu_{\alpha\rho} \tag{10.1}$$

which describes a variation of vectors during the parallel transport [213]. The action for the gravity-particle interaction reads

$$\int dx \mathcal{L} = \frac{1}{8\pi G}\int dx \sqrt{g}R + \int \sqrt{g_{\mu\nu}dx^\mu dx^\nu}, \tag{10.2}$$

where G is the Newton's constant, $R = g^{\mu\nu}R_{\mu\nu}$ with $R_{\mu\nu} = R^\alpha_{\mu\alpha\nu}$. From the action (10.2) one can derive the equations for the metric resulting from the particle motion

$$R_{\mu\nu} - \frac{1}{2}g_{\mu\nu}R = 8\pi G T_{\mu\nu}, \tag{10.3}$$

where $T_{\mu\nu}$ is the particle's energy-momentum tensor. The particle motion is governed by the geodesic equation

$$\frac{d^2 x^\mu}{ds^2} + \Gamma^\mu_{\nu\sigma}\frac{dx^\nu}{ds}\frac{dx^\sigma}{ds} = 0,$$

where ds^2 is the geodesic length. We could describe the motion of a quantum particle in an environment of gravitons in the Heisenberg picture in a way similar to what we did in QED in Sect. 9.6. Then, the geodesic equation would be a counterpart of the Lorentz equation (9.70). In such a case Γ would play the role of the field strength. The Christoffel connection Γ is to be quantized (the quantum geodesic motion is discussed in [20, 143]). However, Γ is not covariant with respect to a change of coordinates. We consider in this chapter a relative motion of a particle of mass m_0 in the falling frame of another heavier particle. The relative motion of these particles is described by the equation of geodesic deviation. Let q^μ be the vector connecting the positions of the particles on adjacent geodesics x^μ, then the acceleration of q^μ satisfies the covariant equation [213]

$$\frac{d^2q^\mu}{ds^2} = R^\mu_{\nu\sigma\rho}u^\nu u^\sigma q^\rho,$$

(10.4)

where $u^\mu = \frac{dx^\mu}{ds}$. The Lagrangian (10.2) as well as the geodesic equation (10.4) depend on the curvature tensor which is covariant with respect to the change of coordinates. This allows a choice of coordinates. We use an assumption of a weak gravitational filed $g_{\mu\nu} = \eta_{\mu\nu} + \lambda h_{\mu\nu}$ (where $\eta_{\mu\nu}$ is the flat Minkowski metric $\mu, \nu = 0, 1, 2, 3$) and $\lambda^2 = 8\pi G$. In the non-relativistic approximation we have (we set the velocity of light $c = 1$) $q = (t, \mathbf{q})$ and $u = (1, \mathbf{0})$. We choose the coordinates for the vacuum Einstein equations leading to the traceless transverse gauge with $h_{\mu 0} = 0$, $\partial_j h_{jk} = 0$ and $h^j_j = 0$, ($j, k = 1, 2, 3$). Then, from Eq. (10.4) in the non-relativistic approximation and in the transverse traceless gauge of the gravitational field we obtain [167, 194, 213] the geodesic equation describing a particle motion in the frame moving with another (reference) particle

$$\frac{d^2q^j}{dt^2} = \frac{1}{2}\lambda\frac{d^2h^{jr}}{dt^2}q^r.$$

(10.5)

10.2 Quantum Geodesic Deviation

The heuristic Feynman prescription for quantization is expressed by the functional integral $\mathcal{D}g \exp(i\int \mathcal{L})$ over the metric. The metric in the Lagrangian (10.2) depends on the choice of coordinates. The transformations of the metric under a change of coordinates can be treated like non-Abelian gauge transformations discussed in the next chapter. In the functional integral quantization of the linearized gravity we have the same problem as in Sect. 9.1 with the electromagnetic field. The operator in the bilinear form in h^{jr} resulting from $\int dx \sqrt{g}R$ is not invertible. In order to quantize the metric we need to choose some special coordinates (gauge). As in non-Abelian gauge theories (Chap. 11) in general the choice of gauge requires an introduction of compensatory (Faddev-Popov) functions in the functional integral which are not Gaussian (for the Faddeev-Popov procedure in gravity see [238]). The transverse-traceless gauge is an analog of the radiation gauge in QED which allows to treat the metric as a Gaussian variable in the lowest order approximation in λ. The Lagrangian (10.2) in the non-relativistic approximation and in the linearized gravity describing the geodesic deviation reads [167, 194] (in Fourier components)

$$\int d\mathbf{x}\mathcal{L} = \frac{1}{8}\int d\mathbf{k}h^*_\alpha(\mathbf{k}, t)((\tfrac{d}{dt})^2 + k^2)h_\alpha(\mathbf{k}, t) + \frac{1}{2}m_0\frac{dq_r}{dt}\frac{dq_r}{dt}$$
$$-\frac{1}{4}m_0\lambda(2\pi)^{-\frac{3}{2}}\int d\mathbf{k}\exp(i\mathbf{kq})\frac{d^2h_{rl}}{dt^2}(\mathbf{k})q_r q_l.$$

(10.6)

In Eq. (10.6) we decompose h_{rl} into the amplitudes h_α (where $\alpha = +, \times$ in the linear polarization) by means of the polarization tensors e^α_{rl} [213]

$$h_{rl}(\mathbf{x}, t) = (2\pi)^{-\frac{3}{2}}\int d\mathbf{k}(h_\alpha(\mathbf{k})e^\alpha_{rl}\exp(-i\mathbf{kx}) + h^*_\alpha(\mathbf{k})e^\alpha_{rl}\exp(i\mathbf{kx})).$$

(10.7)

The quantum gravitational Hamiltonian $H = H_+ + H_\times$ has the same form as the one for two independent scalar fields h_α

$$H = \frac{1}{2} \int d\mathbf{k}\left(-\frac{\delta}{\delta h^\alpha(\mathbf{k})} \frac{\delta}{\delta h^\alpha(-\mathbf{k})} + k^2 h^\alpha(\mathbf{k}) h^\alpha(-\mathbf{k}) \right). \tag{10.8}$$

As in Sect. 8.7 we consider Gaussian states of gravitons which are solutions of the Schrödinger equation ($i\hbar\partial_t \psi = H\psi$)

$$\psi_t^\Gamma(h) = A \exp\left(-\frac{1}{2\hbar} h\Gamma(t)h + \frac{i}{\hbar} J(t)h \right), \tag{10.9}$$

where Γ is an operator (it is unrelated to the Christoffel symbol $\Gamma^\mu_{\sigma\rho}$ of Sect. 10.1) defined by a bilinear form $\Gamma(t, \mathbf{x} - \mathbf{y})$ and $J(t)h = \int d\mathbf{x} J_\alpha(\mathbf{x}) h_\alpha(\mathbf{x})$. Inserting ψ_t^Γ in the Schrödinger equation with the Hamiltonian (10.8) we obtain equations for A, Γ, J (in Fourier space; $\Gamma(\mathbf{k})$ is the Fourier transform of $\Gamma(\mathbf{x})$),

$$\partial_t J = -i\Gamma J, \tag{10.10}$$

$$i\partial_t \Gamma = \Gamma^2 - k^2. \tag{10.11}$$

As in Sect. 8.7 the non-linear (Riccati) equation (10.11) can be related to a linear equation for u defined as

$$u(t) = \exp\left(i \int^t ds \Gamma_s \right). \tag{10.12}$$

Then, u satisfies the equation

$$(\partial_t^2 + k^2)u(k) = 0. \tag{10.13}$$

A solution is (like (8.61) for the scalar field in Sect. 8.7)

$$u = \cos(kt + \alpha - i\nu), \tag{10.14}$$

where the parameters α and ν may depend on k. Then,

$$i\Gamma = -k \tan(kt + \alpha - i\nu) = k\left(i \sinh(2\nu) - \sin(kt + \alpha)\cos(kt + \alpha) \right)$$
$$\times \left(\cosh(2\nu) + \cos(2kt + 2\alpha) \right)^{-1}. \tag{10.15}$$

The probability density is

$$|\psi_t^\Gamma|^2 = \exp\left(-\frac{1}{2\hbar} h(\Gamma + \Gamma^*)h + \frac{i}{\hbar}(J - J^*)h \right), \tag{10.16}$$

where the squeezing is determined by

$$\Gamma(t) + \Gamma^*(t) = 2k \sinh(2\nu)\Big(\cosh(2\nu) + \cos(2kt + 2\alpha)\Big)^{-1}. \qquad (10.17)$$

The fluctuations of h as proportional to $(\Gamma + \Gamma^*)^{-1}$ are large if ν is small (the choice of u in the form (10.14) has the virtue that small Γ corresponds to a small ν). It has been shown that if the gravitational waves are emitted during the inflationary expansion [18, 119] then they are squeezed, i.e., $(\Gamma + \Gamma^*)^{-1}$ is large. In [193, 194] it is suggested that we can detect gravitons by measuring the noise in the squeezed states (in QED the squeezed noise is calculated in Sect. 9.7).

10.3 Heisenberg Equations

We discuss in this section the particle-graviton interaction in a complete analogy with the particle-photon interaction discussed in Chap. 9. From the Lagrangian (10.6) we obtain an equation for the gravitational field

$$\frac{d^2 h^{rl}(\mathbf{k})}{dt^2} + k^2 h^{rl}(\mathbf{k}) = (2\pi)^{-\frac{3}{2}} \lambda f^{rl} \exp(-i\mathbf{k}\mathbf{q}(t)), \qquad (10.18)$$

where

$$f^{rl} = \frac{m_0}{2} \frac{d^2}{dt^2} q^r q^l. \qquad (10.19)$$

We solve Eq. (10.18) for $t \geq t_0$ in the transverse-traceless gauge assuming that when $t \leq t_0$ then h_{rl} is a free wave

$$\begin{aligned} h_{rl}(\mathbf{k}) &\equiv h_{rl}^f + h_{rl}^I \equiv e_{rl}^\alpha (h_0^\alpha \cos(kt) + k^{-1}\Pi_0^\alpha \sin(kt)) \\ &+ \lambda(2\pi)^{-\frac{3}{2}} \Lambda_{rl;mn} \int_{t_0}^t k^{-1} \sin(k(t-t')) f_{mn}(t') \exp(-i\mathbf{k}\mathbf{q}(t'))dt'. \end{aligned} \qquad (10.20)$$

Λ is a projection onto the transverse-traceless gauge

$$\begin{aligned} \Lambda_{ij;mn}(\tfrac{\mathbf{k}}{k}) &= e_{ij}^\alpha e_{mn}^\alpha = \tfrac{1}{2}(\delta_{im} - k^{-2}k_i k_m)(\delta_{jn} - k^{-2}k_j k_n) \\ &+ \tfrac{1}{2}(\delta_{in} - k^{-2}k_i k_n)(\delta_{jm} - k^{-2}k_j k_m) \\ &- \tfrac{1}{3}(\delta_{ij} - k^{-2}k_i k_j)(\delta_{nm} - k^{-2}k_n k_m). \end{aligned} \qquad (10.21)$$

h_{rl}^f is expanded in Fourier modes

$$h_{rl}^f(t, \mathbf{x}) = (2\pi)^{-\frac{3}{2}} \int d\mathbf{k} \exp(i\mathbf{k}\mathbf{x}) h_{rl}^f(\mathbf{k}, t).$$

It is a superposition of quantum free modes (10.7) decomposed into h_0^α and Π_0^α as quantum initial conditions describing the environment of gravitons (an analog of the environment of photons in Eq. (9.80)). h_{rl}^I in Eq. (10.20) is the gravitational field created by the particle motion. h_{rl}^f satisfies the homogeneous equation

$$\frac{d^2 h_{rl}^f(\mathbf{k})}{dt^2} + k^2 h_{rl}^f(\mathbf{k}) = 0.$$

The equation of motion (10.5) resulting from the Lagrangian (10.6) for the coordinate $\mathbf{q}$ after an insertion of the solution (10.20) is

$$\frac{d^2 q_r}{dt^2} = \frac{\lambda}{2}(2\pi)^{-\frac{3}{2}} \int d\mathbf{k} \exp(i\mathbf{k}\mathbf{q}(t)) \frac{d^2 h_{rl}^I(\mathbf{k})}{dt^2} q^l(t)$$
$$+ \frac{\lambda}{2}(2\pi)^{-\frac{3}{2}} \int d\mathbf{k} \exp(i\mathbf{k}\mathbf{q}(t)) \frac{d^2 h_{rl}^f(\mathbf{k})}{dt^2} q^l(t). \tag{10.22}$$

Equation (10.22) can be written in the form

$$\frac{d^2 q_r}{dt^2} = F_r(\mathbf{q}, t) + N_{rl}(t, \mathbf{q}) q^l, \tag{10.23}$$

where F is a non-linear interaction (the backreaction discussed in the next section) resulting from the interaction of the mass m_0 with its own gravitational field and $N_{rl}(t, \mathbf{q})$ is the noise. The noise $N_{rl}(t, \mathbf{q})$ is expressed by the initial values of the canonical (free) variables $h_0^\alpha(\mathbf{k})$ and $\Pi_0^\alpha(\mathbf{k})$

$$N^{rl}(t, \mathbf{q}) = \int d\mathbf{k} N^{rl}(\mathbf{k}, \mathbf{q}, t) = \frac{\lambda}{2}(2\pi)^{-\frac{3}{2}} \int d\mathbf{k} \exp(i\mathbf{k}\mathbf{q}(t)) k^2 e_{rl}^\alpha$$
$$\times (h_0^\alpha \cos(kt) + k^{-1} \Pi_0^\alpha \sin(kt)). \tag{10.24}$$

10.4   Stochastic Motion in the Thermal Environment

The arbitrary free modes h_{rl}^f appear in the solution of particle-gravity equations (10.3)–(10.4) in the same way as in electrodynamics in Sect. 9.6. They describe the effect of the environment upon the particle motion. In electrodynamics there is a well-explored classical environment of a cavity filled with the electromagnetic radiation. It has a quantum version which in the limit $\beta \to \infty$ describes the vacuum photon fluctuations responsible for the Lamb shift. The corresponding effects of gravitons as proportional to the Newton constant G are incomparably weaker [38] as $m_0 G e^{-2} \simeq 10^{-42}$ for the electron. The gravitons in an early stage of universe evolution (before inflation, at the Planck time [237]) can be in an equilibrium with all particles of a unified quantum field theory of all interactions. We discuss in this section the effect of such a thermal environment of gravitons upon a particle motion (the noise in the graviton's squeezed state is discussed in [145]).

In a linearized quantum gravity in a thermal equilibrium at the temperature β^{-1} we have a formula for correlation functions analogous to Eq. (9.81)

$$
\begin{aligned}
< h_{rl}^q(t, \mathbf{x}) h_{mn}^q(t', \mathbf{x}') > &= Tr(\rho h_{rl}^q(t, \mathbf{x}) h_{mn}^q(t', \mathbf{x}')) \\
&= \hbar (2\pi)^{-3} \int d\mathbf{k} k^{-1} \Lambda^{rl;mn} \cos(\mathbf{k}(\mathbf{x} - \mathbf{x}')) \\
&\times \left(\cos(k(t - t')) \coth(\tfrac{1}{2}\beta k \hbar) - i \sin(k(t - t')) \right).
\end{aligned}
\tag{10.25}
$$

Then, the operators $N^{rl}(t, \mathbf{q})$ (10.24) do not commute

$$
\begin{aligned}
[N^{rl}(t, \mathbf{q}), N^{mn}(t', \mathbf{q}')] &= i\hbar \tfrac{\lambda^2}{4} (2\pi)^{-3} \int d\mathbf{k} k^3 \Lambda^{rl;mn} \\
&\times \sin(k(t - t')) \exp(i\mathbf{k}\mathbf{q}(t) - i\mathbf{k}\mathbf{q}(t')).
\end{aligned}
\tag{10.26}
$$

From Eq. (10.25) the quantum thermal Bose-Einstein distribution is

$$
< h^\alpha(\mathbf{k}) h^{\alpha'}(\mathbf{k}') > = \delta^{\alpha\alpha'} \frac{1}{2} \hbar k^{-1} \coth\left(\frac{1}{2} \hbar \beta k \right) \delta(\mathbf{k} + \mathbf{k}')
\tag{10.27}
$$

and

$$
< \Pi^\alpha(\mathbf{k}) \Pi^{\alpha'}(\mathbf{k}') > = \delta^{\alpha\alpha'} \frac{1}{2} \hbar k \coth\left(\frac{1}{2} \beta k \right) \delta(\mathbf{k} + \mathbf{k}').
\tag{10.28}
$$

The correlation functions of the noise are

$$
\frac{1}{2}\left\langle N^{mn}(\mathbf{q}, t') N^{rl}(\mathbf{q}, t) + N^{rl}(\mathbf{q}, t) N^{mn}(\mathbf{q}, t') \right\rangle = \tfrac{\lambda^2}{4}(2\pi)^{-3} \int d\mathbf{k} k^4 \Lambda_{rl;mn}
\times \tfrac{1}{2} \hbar k^{-1} \coth(\tfrac{1}{2}\hbar\beta k)) \cos(k(t - t')) \exp(i\mathbf{k}\mathbf{q}(t') - i\mathbf{k}\mathbf{q}(t)).
\tag{10.29}
$$

Equation (10.22) is a non-linear and non-local integro-differential equation. It simplifies if we assume $\mathbf{k}\mathbf{q} \simeq 0$. We denote $N_{rl}(t) \equiv N_{rl}(t, \mathbf{0})$. Then

$$
\begin{aligned}
N^{rl}(t) &= \int d\mathbf{k} N^{rl}(k, \mathbf{0}, t) \\
&= \tfrac{\lambda}{2}(2\pi)^{-\frac{3}{2}} \int d\mathbf{k} k^2 e_{rl}^\alpha (h_0^\alpha \cos(kt) + k^{-1} \Pi_0^\alpha \sin(kt)).
\end{aligned}
\tag{10.30}
$$

Let us consider some simplifications of Eqs. (10.25)–(10.30). First, we consider the (classical or high temperature) limit $\beta\hbar \to 0$. Then, the field $h^\alpha(\mathbf{k})$ and its canonically conjugated momentum $\Pi^\alpha(\mathbf{k})$ are distributed according to the classical Gibbs distribution

$$
d\Pi_0^\alpha dh_0^\alpha \exp\left(-\frac{\beta}{2} \int d\mathbf{k}(|\Pi_\alpha|^2 + k^2 |h_\alpha|^2) \right).
\tag{10.31}
$$

From Eqs. (10.27)–(10.28) in the limit $\beta\hbar \to 0$

$$
< h_0^\alpha(\mathbf{k}) h_0^{\alpha'}(\mathbf{k}') > = \delta^{\alpha\alpha'} \beta^{-1} k^{-2} \delta(\mathbf{k} + \mathbf{k}')
\tag{10.32}
$$

and

$$
< \Pi_0^\alpha(\mathbf{k}) \Pi_0^{\alpha'}(\mathbf{k}') > = \delta^{\alpha\alpha'} \beta^{-1} \delta(\mathbf{k} + \mathbf{k}').
\tag{10.33}
$$

We calculate the noise (10.24) at $\mathbf{kq} = 0$ and $\beta\hbar \to 0$ as given by Eq. (10.30). It has the correlations

$$\left\langle N^{rl}(t)N^{mn}(t')\right\rangle = < \Lambda_{rl:mn} > \frac{\lambda^2}{4\pi}\beta^{-1}\partial_t^2\partial_{t'}^2\delta(t-t'), \qquad (10.34)$$

where the angular average $< \Lambda >$ of Λ over $\mathbf{k}k^{-1}$ is

$$\frac{1}{4\pi} < \Lambda_{ij;mn} > = \frac{1}{5}(\delta_{im}\delta_{jn} + \delta_{in}\delta_{jm}) - \frac{2}{15}\delta_{ij}\delta_{nm}. \qquad (10.35)$$

If the dependence on the coordinate $\mathbf{q}$ is considered (but we ignore the non-commutativity of $\mathbf{q}$'s) then from Eqs. (10.32)–(10.33)

$$\left\langle N^{rl}(\mathbf{q},t)N^{mn}(\mathbf{q},t')\right\rangle = \frac{\lambda^2}{4}\beta^{-1}(2\pi)^{-3}\int d\mathbf{k}k^2\Lambda_{rl:mn}$$
$$\times \cos(k(t-t'))\exp(i\mathbf{kq}(t') - i\mathbf{kq}(t)). \qquad (10.36)$$

In computation of the noise correlations (10.36) we can first calculate the angular average over $k^{-1}\mathbf{k}$ using the formula

$$\int d\theta d\phi \sin\theta \Lambda^{rl;mn}(\mathbf{k}k^{-1})\exp(i\mathbf{kx}) = 4\pi\Lambda^{rl;mn}(k^{-1}\nabla_{\mathbf{x}})(k|\mathbf{x}|)^{-1}\sin(k|\mathbf{x}|). \qquad (10.37)$$

We can calculate the integral (10.36) performing the k-integral of trigonometric functions. In general, the result will depend on the tensor $(q_r(t) - q_r(t'))(q_l(t) - q_l(t'))|\mathbf{q}(t) - \mathbf{q}(t')|^{-2}$ (such terms are more singular). The term proportional to the tensor (10.35) (the one without the tensorial q -terms) is

$$\left\langle N^{rl}(\mathbf{q},t)N^{mn}(\mathbf{q},t')\right\rangle$$
$$= \frac{5}{2} < \Lambda >_{rl:mn} \frac{\lambda^2}{16\pi}\beta^{-1}|\mathbf{q}(t) - \mathbf{q}(t')|^{-1}\partial_t^2\partial_{t'}^2$$
$$\times \left(\epsilon(t - t' + |\mathbf{q}(t) - \mathbf{q}(t')|) - \epsilon(t - t' - |\mathbf{q}(t) - \mathbf{q}(t')|)\right), \qquad (10.38)$$

where the partial derivatives ∂_t and $\partial_{t'}$ in Eq. (10.38) do not act upon $\mathbf{q}$. In Eq. (10.38) we have used the formula (9.92).

There remains to calculate the non-linear force in Eq. (10.23) resulting from an interaction of the particle with its own gravitational field. We have

$$F_r = \frac{1}{2}\lambda^2(2\pi)^{-3}\int d\mathbf{k}\exp(i\mathbf{kq}(t))q^l(t)$$
$$\Lambda_{rl:mn}\partial_t^2\int_{t_0}^t k^{-1}\sin(k(t-t'))\exp(-i\mathbf{kq}(t'))f_{mn}(t')dt'$$
$$= \frac{1}{2}\lambda^2(2\pi)^{-3}\int d\mathbf{k}\exp(i\mathbf{kq}(t))q^l(t)\Lambda_{rl;mn}$$
$$\times (\int_{t_0}^t dt'\partial_t \cos(k(t-t'))\exp(-i\mathbf{kq}(t')f_{mn}(t') + f_{mn}(t)\exp(-i\mathbf{kq}(t)). \qquad (10.39)$$

We write the last factor in Eq. (10.39) as

$$
\int_{t_0}^{t} \partial_t \cos(k(t-t')) \exp(-i\mathbf{k}\mathbf{q}(t')) f_{mn}(t')dt' + f_{mn}(t) \exp(-i\mathbf{k}\mathbf{q}(t))
$$

$$
= -\int_{t_0}^{t} \partial_{t'} \cos(k(t-t')) \exp(-i\mathbf{k}\mathbf{q}(t')) f_{mn}(t')dt' + f^{mn}(t) \exp(-i\mathbf{k}\mathbf{q}(t))
$$

$$
= \int_{t_0}^{t} dt' \cos(k(t-t')) \partial_{t'} \Big(f_{mn}(t') \exp(-i\mathbf{k}\mathbf{q}(t')) \Big)
$$
$$
+ \cos(k(t-t_0)) f_{mn}(t_0) \exp(-i\mathbf{k}\mathbf{q}(t_0))
$$

$$
= \int_{t_0}^{t} dt' \cos(k(t-t')) \partial_{t'} \Big(f_{mn}(t') \exp(-i\mathbf{k}\mathbf{q}(t')) \Big)
$$

$$
= \int_{t_0}^{t} dt' \cos(k(t-t')) \exp(-i\mathbf{k}\mathbf{q}(t')) (\partial_{t'} f_{mn}(t') - i\mathbf{k}\tfrac{d\mathbf{q}}{dt'} f_{mn}(t)'),
$$

(10.40)

where we assumed $f_{mn}(t_0) = 0$. In the approximation $\mathbf{k}\mathbf{q}(t') \simeq 0$ the integral (10.40) over dk is

$$
\int_{t_0}^{t} \int dk k^2 \cos(k(t-t')) \partial_{t'} f_{mn}(t')dt'
$$

(10.41)

$$
= -2\pi \int_{t_0}^{t} \partial_{t'}^2 \delta(t-t') \partial_{t'} f_{mn}(t')dt' = 2\pi \partial_t^3 f_{mn}(t).
$$

As in the case (9.97) of a particle in an electromagnetic field we treat the operator equation (10.23) semiclassically as an equation for an average $< q(t) >$ in a quantum state of the graviton. Then, from Eqs. (10.39)–(10.41) we can write Eq. (10.23) in the form

$$
\frac{d^2 q^r}{dt^2} = -\frac{\lambda^2}{5\pi}(\delta_{rm}\delta_{ln} - \tfrac{1}{3}\delta_{rl}\delta_{mn})q^l \partial_t^3 f^{mn} + N^{rl}(t)q_l.
$$

(10.42)

This is the perturbation by noise obtained in [144, 194] of the gravitational back-reaction equation of ref. [201]. The noise is necessary in the classical backreaction equation if the particle is to stay in equilibrium with the gravitational radiation in accordance with the fluctuation dissipation theorem of statistical physics. The noise can be strengthened by inflation which is producing the squeezed states of gravitons [18] (for a discussion of the noise of gravitons in squeezed states see [145]). Eq. (10.42) obtained in the approximation of small $\mathbf{q}$ has a simple local form. The correlation functions of the noise $N^{rl}(t)$ in Eq. (10.42) calculated at $\mathbf{q} = 0$ in Eq. (10.34) are quite singular and difficult for a measurement in configuration space (we can measure the frequency spectrum instead). We could consider the noise (10.29) or (10.36) as appropriate for space-time measurements.

If in Eq. (10.22) we take $\exp(-i\mathbf{k}\mathbf{q}(t'))$ into account then we obtain an integro-differential equation. In the calculations (10.39) and (10.40) we have ignored the non-commutativity of $\mathbf{q}(t)$ and $\mathbf{q}(t')$.

10.5   Exercises

10.1 Calculate the Riemannian tensor $R_{\mu\nu\sigma\rho}$ for the linearized gravitational plane wave and show that Eq. (10.4) in the non-relativistic limit gives Eq. (10.5).

10.2 Calculate the correlations (10.29) from Eq. (10.24) and (10.27)–(10.28).

10.3 Compute the quantum correlations (10.29) in the approximation $\mathbf{kq} \simeq 0$.

10.4 Calculate the n-point correlation functions of the graviton noise in the Gaussian state (10.9).

10.5 Solve Eq. (10.42) perurbatively by iteration in order to calculate the deviation from a straight line motion (note that $\partial_t^3 f_{mn}$ is vanishing in the lowest order).

10.6 Estimate the friction and the spectral line width resulting from the quantum deviation equation (10.42).

Chapter 11
Quantization of Non-Abelian Gauge Fields

Abstract The begin of non-Abelian gauge theories originates from the problem of a generalization of the Abelian group of gauge transformations of the electromagnetic potential of Chap. 9. The solution involves a non-linear coupling of gauge field components. The coupling of gauge fields to matter fields together with the Higgs mechanism of mass generation led to the standard (Weinberg-Salam) model of particle physics which is the basis of the contemporary particle physics describing in a unified way electromagnetic, weak and strong interactions. We discuss briefly the mechanism of mass generation in classical gauge theory as it is the basis of applications to the Standard Model. Then, the quantization of the scalar field in an external gauge field is discussed. The quantization of gauge fields requires the gauge fixing which is achieved in the Faddeev-Popov formalism. The renormalization of gauge theories is discussed at one-loop level. We calculate the one loop effective action in the theory with scalar and gauge fields. The one-loop formula involves the heat kernel representation of differential operators discussed earlier in Chap. 5. We discuss the stability of non-Abelian gauge theories (asymptotic freedom) on the basis of one-loop effective action. A mathematically precise gauge fixing and perturbation expansion in gauge theories is possible after lattice regularization which is discussed in Chap. 12.

The begin of non-Abelian gauge theories originates from the problem of a generalization of the Abelian group of gauge transformations of the electromagnetic potential of Chap. 9 [247]. The solution involves a non-linear coupling of gauge field components. The coupling of gauge fields to matter fields together with the Higgs mechanism of mass generation led to the standard (Weinberg-Salam) model of particle physics which is the basis of the contemporary particle physics describing in a unified way electromagnetic, weak and strong interactions. The quantization and renormalization of non-Abelian gauge theories encountered severe difficulties which have been overcome by t'Hooft and Veltman [4, 232, 233]. The discovery of the asymptotic freedom [120, 197] explained the free behavior of quarks (partons) in strong interactions and became a basis for a description of hadrons as confined

© The Author(s), under exclusive license to Springer Nature Switzerland AG 2023   189
Z. Haba, *Lectures on Quantum Field Theory and Functional Integration*,
https://doi.org/10.1007/978-3-031-30712-6_11

quarks in quantum chromodynamics (QCD). In this chapter we discuss only the
bosonic part of these interactions. Sect. 11.1 is an introduction to classical gauge
theory. In Sect. 11.2 we discuss briefly the mechanism of symmetry breaking and
mass generation in classical gauge theory as it is the basis of applications to the
Standard Model. The quantization of the scalar field in an external gauge field is
discussed in Sect. 11.3. The gauge fixing (Faddeev-Popov procedure) necessary for
quantization of the gauge field is studied in Sect. 11.4. In Sect. 11.5 the one loop
effective action in the theory with scalar and gauge fields is calculated.The one-loop
formula involves the heat kernel representation of differential operators discussed
earlier in Chap. 5. A mathematically precise gauge fixing and perturbation expansion
in gauge theories is possible after lattice regularization which is discussed in the next
chapter.

11.1 Non-Abelian Gauge Theories

Let us introduce a matrix-valued gauge potential

$$A_\mu = \tau_a A_\mu^a, \tag{11.1}$$

where τ^a are Hermitian matrices with the commutation relations (as a representation
of the algebra of a compact group $\mathcal{G}$, usually $SU(n)$ or $O(n)$; the term unitary should
be replaced by orthogonal in the latter case)

$$[\tau_a, \tau_b] = i f_{abc} \tau_c, \tag{11.2}$$

where f_{abc} are the structure constants of the Lie algebra of the group (Sect. 1.3). The
gauge transformation is defined by a unitary matrix U

$$A_\mu^U = U^{-1} A_\mu U + i U^{-1} \partial_\mu U. \tag{11.3}$$

The transformation (11.3) is expressed by a unitary representation U of the group
$\mathcal{G}$ but the transformation of A_μ^a does not depend on the representation. This can be
seen from its infinitesimal form (where $U = 1 - i\tau^a u^a$)

$$A_\mu^{a'} = A_\mu^a + f^{abc} A_\mu^b u^c + \partial_\mu u^a.$$

We could say that the transformation is expressed in the adjoint representation τ_{ad}
whose matrix elements are defined by the structure constants

$$(\tau_{ad}^r)_{bc} = i f_{rbc}.$$

We shall also denote the transformation as A^G ($G \in \mathcal{G}$) in order to emphasize that
the transformation of A does not depend on the representation of $\mathcal{G}$. The Lagrangian

is (μ_0 appearing in QED in Chap. 9 is absorbed in the coupling constant g^2)

$$\mathcal{L} = -\frac{1}{4g^2}Tr(F_{\mu\nu}F^{\mu\nu}) = -\frac{1}{4g^2}F^a_{\mu\nu}F^{a\mu\nu}, \tag{11.4}$$

where $F_{\mu\nu} = F^a_{\mu\nu}\tau^a$ with

$$F_{\mu\nu} = \partial_\mu A_\nu - \partial_\nu A_\mu + i[A_\mu, A_\nu]. \tag{11.5}$$

It can be shown that under the gauge transformation (11.3) $F_{\mu\nu} \to U^{-1}F_{\mu\nu}U$. The standard perturbation theory involves a change of variables by a scaling of potentials by means of the coupling constant g

$$A = g\tilde{A}. \tag{11.6}$$

Then, at $g = 0$ the Lagrangian (11.4) is quadratic in $\tilde{A}$. We need n-dimensional representations U of the group $\mathcal{G}$ in order to define its action $\phi \to U\phi$ upon the n-tuple of the scalar field ϕ. We expand the Lagrangian in g (with the scaled fields $\tilde{A}$). Then, at the lowest order we encounter the same problem as in the Abelian case in Sect. 9.1 that the operator M is not invertible. The physicists in the sixties tried to solve the problem in the same way as in the Abelian case: introducing the gauge fixing like the one in Eq. (9.6). However, after the calculation of the scattering amplitudes they recognized that such a procedure violated unitarity. We are not allowed to introduce ad hoc gauge fixing term without further modification of the functional integral.

   The important question in theories with a gauge invariance concerns the identification of variables covariant and invariant with respect to the gauge transformations. An important covariant variable is a line itegral [169]

$$U^A(y, x) = T\left(\exp\left(i\int_x^y A_\mu d\xi^\mu\right)\right) \tag{11.7}$$

integrated along a path ξ (product integral [70]).

   It follows from its definition that $U^A(y, x)$ satisfies the equation

$$dU^A(\xi(\tau), x) = i A^\mu(\xi(\tau))U^A(\xi(\tau), x)d\xi_\mu(\tau). \tag{11.8}$$

The solution of Eq. (11.8) can also be considered as a definition of the product integral (11.7). It can be shown using Eqs. (11.3) and (11.8) that

$$U^{A^U}(y, x) = U^{-1}(y)U^A(y, x)U(x). \tag{11.9}$$

In order to prove Eq. (11.9) we differentiate both sides of Eq. (11.9) over y using (11.8) and show that lhs satisfies Eq. (11.8) with a gauge-transformed vector potential on the rhs. In detail,

$$dU^{A^U}(y,x) = dU^{-1}(y)U^A(y,x)U(x) + U^{-1}(y)dU^A(y,x)U(x)$$
$$= -U^{-1}dUU^A(y,x)U(x) + U^{-1}(y)iA_\mu d\xi^\mu U^A(y,x)U(x)$$
$$= \left(iU^{-1}(y)A_\mu d\xi^\mu U(y) + U^{-1}(y)dU(y)\right)U^{-1}(y)U^A(y,x)U(x)$$

It follows from Eq. (11.9) that when calculated along a closed curve $Tr(U^{A^U}(x,x)) = Tr(U^{-1}(x)U^A(x,x)U(x)) = Tr(U^A(x,x))$. So, $Tr(U^A(x,x))$ is gauge invariant.

11.2 The Non-Abelian Higgs Model: Symmetry Breaking and Mass Generation

The non-Abelian Higgs model is a generalization of the one of Sect. 9.2 defined by the Lagrangian in Eq. (9.16)

$$\mathcal{L} = \frac{1}{2}(\partial_\mu \phi^a)^* \partial^\mu \phi^a - \frac{1}{2}m^2(\phi^a)^*\phi^a - \frac{\lambda^2}{8}((\phi^a)^*\phi^a)^2.$$

We represent the n-tuple of scalar fields ϕ^a, a=1,...,n, by a column. We consider the n-dimensional representation of the group $\mathcal{G}$ (real orthogonal if the fields are real or complex unitary if the fields are complex). We couple the scalar field to the gauge field as (+ means the Hermitian conjugation of a matrix or an operator)

$$\mathcal{L} = \tfrac{1}{2}((\partial^\mu + iA^\mu)\phi)^+(\partial_\mu + iA_\mu)\phi - \tfrac{1}{2}m^2\phi^+\phi - \tfrac{\lambda^2}{8}(\phi^+\phi)^2. \qquad (11.10)$$

It is invariant under the representation U of the group $\mathcal{G}$ acting as $\phi \to U\phi$ on the n-tuple of fields ϕ together with the gauge transformation (11.3).

In the Higgs model $m^2 = -\mu^2 < 0$. Then, the scalar field gains a mass through a non-zero expectation value $v = \pm|v|$ of ϕ at the minimum of the potential $|v| = \frac{\sqrt{2}\mu}{\lambda}$. This minimum v can be achieved in many ways. The simplest one is when a real field ϕ transforms under the fundamental representation of the group $O(n)$ and $v = (v^1, 0, ..., 0)$. Expressing the fields $\phi = \tilde{\phi} + v$ in the Lagrangian (11.10) in terms of the fields $\tilde{\phi}$ we notice from the quadratic part of the Lagrangian (11.10) that the field $\tilde{\phi}^1$ acquires the mass μ and the gauge field $\tilde{A}_\nu^1$ the mass $g|v|$. This is called the Higgs mechanism [153]. The Lagrangian expressed by $\tilde{\phi}$ is not invariant with respect to $O(n)$ group acting upon $\tilde{\phi}$ as $\tilde{\phi} \to U\tilde{\phi}$. Only the invariance with respect to $O(n-1)$ is preserved. The simple scheme of mass generation works well in classical gauge theories [163] in the sense that the solutions of classical field equations decay exponentially in space with a decay rate proportional to the inverse of the mass. In quantum field theory the proof of the mass generation is not simple. It has been proved on the lattice [192] at strong coupling (see Sect. 12.3). The naive argument applied in this section does not work in quantum mechanics (one dimensional field theory) as there is a tunnelling between the two classical vacua.

The correlation functions after the gauge fixing (9.5)

$$\frac{1}{2\xi}(\partial^\mu \tilde{A}^a_\mu)^2$$

are expressed similarly as in the Abelian case of Sect. 9.2. The scalar field gains an index a. The quadratic part in ϕ of the Lagrangian (11.10) defines the propagator for the scalar field. The quadratic part of the Lagrangian (11.4) defines the propagator for each component of the gauge field. The basic difference between Abelian and non-Abelian gauge theories appears in the vector fields self-interaction which is not quadratic. It does not allow the polymer representation of Sect. 9.3 (as we cannot perform the Gaussian integration over the gauge fields). The non-linear vector field interaction constitutes the main difficulty in the proof of renormalizability of the model. In Sect. 11.6 we show renormalizability of the non-Abelian gauge theories only at the level of one-loop (for general proof see [4, 232, 233]).

11.3   The Effective Scalar Field Action in Non-Abelian Gauge Field

In this section we work in the Euclidean framework. The quadratic part of the scalar field Lagrangian (11.10) is defined by the operator

$$\Delta_A = (\partial_\mu + iA_\mu)^+(\partial_\mu + iA_\mu). \tag{11.11}$$

In order to express the Green's functions of this operator let us consider a solution of the differential equation (11.8) with a Brownian path $w^\mu(\tau)$

$$dT^A_\tau = iA^\mu(w_x)T^A_\tau \circ dw^\mu_x = \tfrac{1}{2}(i\partial_\mu A_\mu - A_\mu A_\mu)T^A_\tau d\tau + iA^\mu(w_x)T^A_\tau dw^\mu_x, \tag{11.12}$$

where $w^\mu_x = x^\mu + w^\mu$ and we applied the relation between Stratonovitch and Ito integrals (5.62). We prove that

$$(T^A_\tau)^+ T^A_\tau = 1 \tag{11.13}$$

taking the Stratonovitch differential of Eq. (11.13) and proving that the differential of both sides of Eq. (11.13) is zero (the constant on the rhs of Eq. (11.13) must be 1 because $T^A_0 = 1$)

$$(dT^A_\tau)^+ \circ T^A_\tau + (T^A_\tau)^+ \circ dT^A_\tau = (T^A_\tau)^+(-iA_\mu \circ dw_\mu)T^A_\tau + (T^A_\tau)^+ iA_\mu \circ dw_\mu T^A_\tau = 0.$$

The solution T^A_τ is gauge covariant: if

$$A^U_\mu = U^+ A_\mu U + iU^+ \partial_\mu U \tag{11.14}$$

then

$$T^{A^U}_\tau = U^+(w(\tau))T^A_\tau U(x). \tag{11.15}$$

Similarly as for Eq. (11.9) it is sufficient to prove that the Stratonovitch differential of both sides of Eq. (11.15) is the same (Eq. (11.15) shows that the Stratonovitch differential dT_τ^A (11.12) has the same transformation and differentiation properties as $U^A(y, x)$ in Eq. (11.9)). Differentiating the rhs of Eq. (11.15) we get

$$dT_\tau^{A^U} = dU^+(w(\tau)) \circ T_\tau^A U(x) + U^+(w(\tau)) \circ dT_\tau^A U(x).$$

Then, using (11.12) for dT_τ^A and $dU = \partial_\mu U \circ dw_\mu$ on the rhs of this equation we convince ourselves that the rhs is equal to the differential of $dT_\tau^{A^U}$ (as in the proof of Eq. (11.9)).

The Green function of the operator $\triangle_A$ is now expressed in the form (an analog of (9.26))

$$(-\triangle_A + m^2)^{-1}(x, y) = \frac{1}{2} \int_0^\infty d\tau d W_{(x,y)}^\tau(w) \exp\left(-\frac{1}{2}m^2\tau\right) T_\tau^A. \qquad (11.16)$$

Note that from $|T_\tau^A| = 1$, where $|..|$ is the norm of the matrix, there follows the diamagnetic inequality [211]

$$|(-\triangle_A + m^2)^{-1}(x, y)| \le \frac{1}{2} \int_0^\infty d\tau d W_{(x,y)}^\tau(w) \exp(-\frac{1}{2}m^2\tau)$$
$$= (-\triangle + m^2)^{-1}(x, y)$$

saying that the propagator in a gauge field is bounded by the propagator without the gauge field.

In order to prove Eq. (11.16) it is sufficient to show that for the time differentials we have the equality

$$d \exp\left(\frac{\tau}{2}\triangle_A\right)(x, y) = d \int d W_{(x,y)}^\tau(w) T_\tau^A = \frac{1}{2}\triangle_A \int d W_{(x,y)}^\tau(w) T_\tau^A d\tau \quad (11.17)$$

as Eq. (11.16) follows by integration over the interval $[0, \infty]$ of Eq. (11.17).
Eq. (11.17) can be shown in the same way as the Feynman integral formula for an evolution in an electromagnetic field in Sect. 5.4, Eq. (5.65) (it is sufficient to apply the differential (11.12) of T_τ^A). The effective action in the one loop approximation is expressed as in Sects. 7.5 and 7.6 by the determinant of the differential operator which subsequently can be represented by the trace of the heat kernel (see Eq. (9.26))

$$Tr\left(\exp\left(-\frac{\tau}{2}\triangle_A\right)\right) = Tr^U \int dx \int d W_{(x,x)}^\tau(w) T_\tau^A, \qquad (11.18)$$

where Tr^U is the trace of the matrix T_τ^A in the U-representation (from now on we begin to distinguish in notation the trace of matrices and the trace of operators denoted simply by Tr).

We can use Eq. (11.18) to calculate the vacuum polarization coming from the interaction with scalar fields at one loop (in the Euclidean version). The vacuum polarization is the coefficient function of the quadratic part $\Gamma_2(A)$ in A in the expansion of the effective action $\Gamma(A) \simeq \frac{1}{2} \ln \det(-\Delta_A + m^2)$ (see Eq. (7.52) in Sect. 7.5). The quadratic part of the action $\frac{1}{4} F^2 + \Gamma_2(A)$ defines a quantum correction to the inverse of the the propagator resulting from the interaction of the scalar field with the gauge field. Using the formula (11.12) we obtain by an expansion of T_t^A in the effective action (9.33) (at one loop) till the second order in A (as Γ_2 is gauge invariant we may choose the Lorentz gauge, $\partial_\mu A_\mu = 0$ as a convenient gauge for calculations)

$$
\begin{aligned}
\Gamma_2(A) &= \tfrac{1}{4} \int_\Omega dx \int_0^\infty \tfrac{dt}{t} \exp(-\tfrac{1}{2}m^2 t) dW_{(x,x)}^t(w) \\
&\times \Big(Tr^U \Big[-\tfrac{1}{2} \int_0^t A_\mu(w(s)) A_\mu(w(s)) ds \\
&- \int_0^t A^\mu(w(s)) dw^\mu(s) \int_0^s A^\nu(w(s')) dw^\nu(s') \Big] \Big) \\
&\equiv \int dp dk A_\mu^a(p) A_\nu^a(k) \delta_\Omega(p,k) \Pi_{\mu\nu}(p,k),
\end{aligned}
\tag{11.19}
$$

where

$$
A(w(s)) = (2\pi)^{-2} \int dp \exp(ipw(s)) A(p),
$$

$$
\delta_\Omega(p,k) = (2\pi)^{-4} \int_\Omega dx \exp(i(p+k)x)
\tag{11.20}
$$

and Ω is the volume cutoff. It is convenient for calculations to use the representation (7.65) and (7.67) $(w(s) \to q(s))$ for closed paths in terms of the unconstrained Brownian motion (5.31) (similarly as in the calculations in Sect. 9.4). Using the Fourier representation of $A(w(s))$ we obtain from Eq. (11.19)

$$
\begin{aligned}
&\Pi_{\mu\nu}(p,k) \\
&= \tfrac{1}{4} \int_0^\infty \tfrac{dt}{t} \exp(-\tfrac{1}{2}m^2 t)(2\pi t)^{-2} E\Big[-\tfrac{1}{2}\delta_{\mu\nu} \int_0^t \exp(i(p+k)(q(s)-x)) ds \\
&- \int_0^t \Big(1 + (\exp(ip(q(s)-x)) - 1)\Big) dq_\mu(s) \\
&\times \int_0^s \Big(1 + (\exp(ik(q(s')-x)) - 1)\Big) dq_\nu(s')\Big].
\end{aligned}
\tag{11.21}
$$

In the polarization tensor (11.19) similarly as in the Abelian effective action calculations in Eq. (9.42) we must subtract the divergent part. For this purpose we have decomposed $\exp(ip(q-x)) = 1 + (\exp(ip(q-x)) - 1)$ in Eq. (11.21) into the term independent of t (which seems to lead to the most divergent integral at small t in Eq. (11.19)) and the second component which is of order $\sqrt{t}$ because $q(s) - x \simeq \sqrt{t}$ for $q(t) = q(0) = x$ (see Eq. (7.65)). The decomposition in Eq. (11.21) corresponds to the decomposition in Eq. (11.19) $A(w(s)) = A(x) + (A(w(s)) - A(x))$. The $A(x)$ term would describe the gauge field mass renormalization. It is equal

$$\tfrac{1}{4} \int_\Omega dx \int_0^\infty \tfrac{dt}{t} (2\pi t)^{-2} \exp(-\tfrac{1}{2} m^2 t)$$

$$\times \check{E}\Big[Tr^U \Big[-\tfrac{1}{2} \int_0^t A_\mu(x)) A_\mu(x) ds - \int_0^t A^\mu(x) dq^\mu(s) \int_0^s A^\nu(x) dq^\nu(s') \Big] \Big]$$

$$(11.22)$$

We have $\int_0^s dq^\nu(s') = q^\nu(s) - x$. Using the Ito formula

$$d(q^\mu(s) q^\nu(s)) = dq^\mu(s) q^\nu(s) + dq^\nu(s) q^\mu(s) + \delta^{\mu\nu} ds$$

and $\int_0^t d(q^\mu q^\nu) = 0$ for closed paths we express the stochastic integral $dq^\mu \int_0^s dq^\nu$ in Eq. (11.22) by an ordinary integral and conclude that the mass renormalization term (11.22) is zero. As explained in Sect. 7.6 the standard perturbation theory is equivalent to the calculation of the determinant $\det(-\triangle_A + m^2) = \det(\partial_\mu \partial_\mu + m^2 + R)$ as $\det(1 + G_E R)$ in an expansion in R (with a certain operator R quadratic in A). Such computations with a regularized propagator G_E usually lead to an appearance of a photon mass renormalization (violating the gauge invariance).

$\Pi_{\mu\nu}(p, q)$ is gauge independent hence it satisfies the relation

$$\Pi_{\mu\nu}(p, q) = (\delta_{\mu\sigma} - p^{-2} p_\mu p_\sigma)(\delta_{\nu\rho} - q^{-2} q_\nu q_\rho) \Pi_{\sigma\rho}(p, q). \qquad (11.23)$$

It follows that $\Pi_{\mu\nu}$ (when $\Omega \to \infty$ in Eq. (11.19), $q = -p$) is of the form

$$\Pi_{\mu\nu}(p, -p) = (\delta_{\mu\nu} - p^{-2} p_\mu p_\nu) \Pi(p^2), \qquad (11.24)$$

A computation of the expectation value (11.21) gives

$$\Pi(p^2) = \tfrac{1}{4} (2\pi)^{-2} \int_0^\infty dt \, t^{-2} \exp(-\tfrac{1}{2} m^2 t)$$
$$\times \int_0^1 ds' \int_0^{s'} ds \Big(1 - \exp(-\tfrac{1}{2} p^2 s(1 - s)) \Big). \qquad (11.25)$$

In order to calculate the expectation value (11.21) we express q by the unconstrained Brownian motion (7.65) and (7.67) and apply the formula

$$E\Big[w_\mu(s_1) w_\nu(s_2) \exp(i \int h_\mu(s) w_\mu(s) ds) \Big]$$
$$= \Big(\delta_{\mu\nu} min(s_1, s_2) - \int h_\mu(s) min(s, s_1) ds \int h_\nu(s') min(s', s_2) ds' \Big)$$
$$\times \exp\Big(-\tfrac{1}{2} \int ds ds' h_\mu(s) h_\mu(s') min(s, s') \Big)$$

In Eq. (11.25) we still must subtract the term

$$\int_0^\infty dt \, t^{-2} \exp\Big(-\frac{1}{2} m^2 t \Big) \int_0^1 ds' \int_0^{s'} ds \frac{t}{2} p^2 s(1 - s)).$$

which in Eq. (11.19) is proportional to $A_\mu(p)(p^2\delta_{\mu\nu} - p_\mu p_\nu)A_\nu(p)$. This is the logarithmic charge renormalization which we encountered in the calculation of the effective action in the Abelian case (see the discussion below Eq. (9.42)).

We have calculated in Eq. (11.19) the quadratic in A term of the expansion of the one-loop effective action. In order to calculate this action exactly we need to compute the trace of the (matrix) heat kernel $\exp(\tau\Delta_A)(x, x)$ as a functional of A. For such calculations it is useful to apply the coordinate gauge [73] around each point x (the trace of the heat kernel does not depend on the choice of gauge)

$$y_\mu A_\mu(x + y) = 0. \tag{11.26}$$

In the coordinate gauge we can express the gauge potential in terms of the field strength

$$A_\mu(x + y) = \int_0^1 d\alpha F_{\nu\mu}(x + \alpha y)\alpha y_\nu. \tag{11.27}$$

Using the expression (11.27) of $A_\mu(x + \alpha w(s))$ in terms of $F_{\mu\nu}$ we can detect in the expansion of $Tr^U(T_\tau^A - 1)$ in powers of τ the lowest order terms by iterating the equation (11.12) for T_τ^A till the second order in A. We can now proceed similarly as in Sect. 9.4 with $y = \sqrt{\tau}r$ in Eq. (11.27). We can pick the terms in the formula for the determinant

$$\int dx \int \tfrac{d\tau}{\tau}(2\pi\tau)^{-2} \exp(-\tfrac{1}{2}m^2\tau)(T_\tau^A - 1),$$

divergent for small τ.

Calculation of the expectation values over the Brownian bridge of these quadratic terms shows that the expression

$$Tr^U\left(T_\tau^A - 1 + \frac{\tau^2}{48}F_{\mu\nu}F_{\mu\nu}\right) \simeq \tau^3 \tag{11.28}$$

for a small τ. It follows that after the subtraction of the logarithmically divergent term

$$\int_0^\infty \frac{d\tau}{\tau} \exp(-\frac{1}{2}m^2\tau)\frac{1}{48}Tr^U(F_{\mu\nu}F_{\mu\nu}) \tag{11.29}$$

the formula for the determinant is finite.

We can express $A_\mu(x + y)$ as

$$A_\mu(x + y) = \frac{1}{2}F_{\nu\mu}(x)y_\nu + l(U^+DFU) \tag{11.30}$$

where l is a certain linear functional, U is a unitary matrix and $D_\alpha F_{\mu\nu} = \partial_\alpha F_{\mu\nu} - i[A_\alpha, F_{\mu\nu}]$ is the covariant derivative.

It follows from the decomposition (11.30) and the argument used at Eq. (9.42) that the determinant is a product of a functional of F and a certain quadratically bounded functional of DF similarly as in the Abelian case in Sect. 9.4. It can further be shown that $-\ln \det M_A^{\frac{1}{2}}$ which is equal to the one-loop scalar effective action behaves as $-K \int dx F^2 \ln F^2 + (l(F, DF))^2)$ for large F with a constant $K > 0$ [135] (where $l(F, DF)$ is linearly bounded in F). Such a behavior means instability as the scalar effective action is unbounded from below (similarly as the one in Abelian case in Sect. 9.4). As we show in Sect. 11.6 owing to the contribution to the effective action resulting from the gauge field self-interaction there is $\tilde{K} \int dx F^2 \ln F^2 + (l(F, DF))^2)$ contribution with a positive sign of $\tilde{K}$ which can make the total effective action for non-Abelian gauge fields bounded from below.

11.4 Fadeev-Popov Procedure

In Sect. 9.1 we introduced an ad hoc gauge fixing for QED claiming that it does not change gauge invariant correlation functions. We prove this claim in this section in the context of general gauge transformations. The argument is not rigorous. It involves functional integration with respect to a heuristic Lebesgue measure on a space of functions and the Haar measure on a continuous product of groups. The proof becomes mathematically precise when the space-time points form a discrete set (a lattice) and the derivatives are replaced by lattice derivatives (this will be discussed in the next chapter). We consider a gauge transformation generated by $U(x) \in \mathcal{G}$ (we denote in this section the group elements by the same letter as group representations acting upon the scalar fields). There is an invariant measure μ on $\prod_x \mathcal{G}(x)$ (the Haar measure on the continuous product of groups $\mathcal{G}$ indexed by space-time points x). Consider a function $\Phi(A)$ (gauge fixing) which is not invariant under the gauge transformation and define $\Delta(A)$ by

$$\Delta(A) \int dU \Phi(A^U) = 1, \qquad (11.31)$$

where $dU = \prod_x d\mathcal{G}(x)$ is the Haar measure on $\prod_x \mathcal{G}(x)$. It has the property $dU = d(UV) = d(VU) = dU^{-1}$ (for any $V \in \prod_x \mathcal{G}(x)$, V is acting as $V(x)U(x)$ at each x) and $\int dU = 1$.

We have

$$\Delta(A) = \Delta(A^U) \qquad (11.32)$$

as from Eq. (11.31)

$$1 = \Delta(A^V) \int dU \Phi(A^{VU}) = \Delta(A^V) \int d(VU) \Phi(A^{VU}) = \Delta(A^V)(\Delta(A))^{-1}$$

$$(11.33)$$

by a change of variables $VU = U'$. We insert the identity (11.31) in the functional integral of a gauge invariant functional $\mathcal{F}(A)$ ($\mathcal{L}(A)$ is a gauge invariant Lagrangian)

$$
\begin{aligned}
< \mathcal{F} > &= \int dA \exp\left(- \int \mathcal{L}(A) \right) \mathcal{F}(A) \\
&= \int dA \exp\left(- \int \mathcal{L}(A) \right) \mathcal{F}(A) \Delta(A) \int dU \, \Phi(A^U).
\end{aligned}
\tag{11.34}
$$

Changing coordinates $A \to A^U$ and using the gauge invariance of $\Delta(A), dA, \mathcal{F}(A)$ and $\mathcal{L}$ we obtain

$$
< \mathcal{F} > = \int dU \int dA \exp\left(- \int \mathcal{L}(A) \right) \Delta(A) \Phi(A) \mathcal{F}(A),
\tag{11.35}
$$

where $\int dU = 1$. Eq. (11.35) shows that we can insert in the functional integral a gauge fixing function $\Phi(A)$ (as in Sect. 9.2) together with the multiplier $\Delta(A)$.

We prove next that expectation values of gauge invariant variables do not depend on the choice of the gauge fixing Φ. In fact, let us take another gauge fixing Φ'

$$
\Delta'(A) \int dU \Phi'(A^U) = 1.
\tag{11.36}
$$

We insert the identity (11.36) into (11.35). We obtain

$$
\begin{aligned}
< \mathcal{F} > &= \int dA \exp\left(- \int \mathcal{L}(A) \right) \Delta(A) \Phi(A) \mathcal{F}(A) \Delta'(A) \int \Phi'(A^U) dU \\
&= \int dA \exp\left(- \int \mathcal{L}(A) \right) \int dV \Delta(A) \Phi(A^V) \mathcal{F}(A) \Delta'(A) \Phi'(A) \\
&= \int dA \exp\left(- \int \mathcal{L}(A) \right) \mathcal{F}(A) \Delta'(A) \Phi'(A),
\end{aligned}
\tag{11.37}
$$

where we changed variables $A^U \to A$, denoted $U^{-1} = V$, used $dV = dU$, the gauge invariance of $\Delta(A)$ and $\Delta'(A)$ and finally applied the definition (11.31). It can be seen that in Eqs. (11.31)–(11.37) only the gauge invariance of the heuristic measure dA needs a justification. In the next step we choose the gauge fixing $\Phi(A)$ which will need formal manipulations whose rigorous meaning needs a lattice approximation (Chap. 12).

Let us choose the ξ-gauge (9.5) expressed as

$$
\Phi(A) = \int dC(x) \exp\left(- \frac{1}{2\xi g^2} \int dx Tr(C^2) \right) \delta(\partial^\mu A_\mu - C),
\tag{11.38}
$$

where the trace is over the adjoint representation.

In formulas like (11.38) it is understood that the integral $dC(x)$ is a product integral of Lebesgue measures over a continuum of points x. A rigorous version of such expressions is possible on the lattice (Chap. 12) when the points x are from a discrete set and the derivative has the meaning of the lattice derivative.

For infinitesimal transformations $U \simeq 1 - iu$, where u belongs to the algebra of the group, we have

$$U^{-1}A_\mu U + iU^{-1}\partial_\mu U \simeq A_\mu + i[u, A_\mu] + \partial_\mu u \equiv A_\mu + D_\mu(A)u. \qquad (11.39)$$

We calculate the term $\triangle(A)$ from the equation (11.31) (only small u contribute)

$$\triangle(A)\int du\delta(\partial^\mu D_\mu(A)u + \partial^\mu A_\mu - C) = 1. \qquad (11.40)$$

We apply to Eq. (11.40) a formal infinite dimensional version of a finite dimensional formula (true if M is invertible, if not then there is the problem of Gribov copies [117])

$$|\det M|\delta(M\mathbf{x} - \mathbf{y}) = \delta(\mathbf{x} - M^{-1}\mathbf{y}) \qquad (11.41)$$

which can be proved by integration of both sides of this equation

$$\int d\mathbf{x}|\det M|\delta(M\mathbf{x} - \mathbf{y})f(\mathbf{x}) = \int d\mathbf{x}\delta(\mathbf{x} - M^{-1}\mathbf{y})f(\mathbf{x}) = f(M^{-1}\mathbf{y}). \qquad (11.42)$$

Using the formula (11.41) we conclude that $\triangle(A)$ does not depend on C and

$$\triangle(A) = |\det \partial^\mu D_\mu(A)|. \qquad (11.43)$$

The operator inside det has matrix elements (acting on ψ vectors as $\partial^\mu A_\mu\psi = A_\mu\partial^\mu\psi + (\partial^\mu A_\mu)\psi$)

$$\partial_\mu\partial^\mu\delta^{ab} - f^{abc}\partial^\mu A_\mu^c. \qquad (11.44)$$

The determinant (11.43) can be expressed by a functional integral in the form

$$\det \partial^\mu D_\mu(A) = \prod_x d\zeta(x)d\overline{\zeta}(x)\exp\left(-\int dx Tr(\partial^\mu\overline{\zeta}D_\mu(A)\zeta)\right) \qquad (11.45)$$

where ζ^a are the anticommuting Faddeev-Popov ghosts [79].

The rules of calculating such integrals are (for discrete variables $\zeta_j = \zeta(x_j)$, independent at every point x_j)

$$\int d\zeta_j^a\zeta_j^a = 1$$

$$\int d\overline{\zeta}_j^a\overline{\zeta}_j^a = 1$$

and $\int d\zeta\overline{\zeta} = \int d\overline{\zeta}1 = \int d\zeta 1 = 0$. One can easily check (for one variable) that

$$\int d\bar{\zeta} d\zeta \exp(\alpha\bar{\zeta}\zeta) = \alpha.$$

We obtain the determinant (11.43) (without the absolute value) integrating the formula (11.45) after a formal diagonalization of the differential operator $\partial^\mu D_\mu$.

We can write the expression in the path integral as

$$
\begin{aligned}
< \mathcal{F} > \int dA \exp\left(-\int dx \mathcal{L}(A)\right) \Phi(A)\Delta(A)\mathcal{F}(A) \\
= \int dC(x) \int dA \exp\left(-\int dx \mathcal{L}(A)\right) \Delta(A)\mathcal{F}(A) \\
\times \exp\left(-\tfrac{1}{2\xi}\int dx Tr(C^2)\right)\delta(\partial^\mu A_\mu - C).
\end{aligned}
\tag{11.46}
$$

The δ-function integrated with $\exp\left(-\tfrac{1}{2\xi}\int dx Tr(C^2)\right)$ is producing the ξ-gauge fixing introduced for the first time in Sect. 9.1. Hence, performing the C-integral in Eq. (11.46) and inserting $\Delta(A)$ as calculated in Eq. (11.43) we obtain finally

$$
\begin{aligned}
< \mathcal{F} > = \int dA \exp\left(-\int dx \mathcal{L}(A)\right)|\det \partial^\mu D_\mu(A)| \\
\times \exp\left(-\tfrac{1}{2\xi}\int dx Tr((\partial^\mu A_\mu)^2)\right)\mathcal{F}(A).
\end{aligned}
\tag{11.47}
$$

From Eq. (11.47) together with (11.45) we obtain the Feynman diagrams applied in perturbative calculations. The result is expressed in terms of propagators connecting the vertices numbering the order of the perturbation expansion (see [53, 80, 196]). We do not need the anticommuting fields (ghosts) for an expansion of the determinant in powers of A. The determinants can be calculated as in Sect. 7.6

$$\det \partial^\mu D_\mu(A) \simeq \det(1 + K) = \exp\left(TrK - \frac{1}{2}Tr(K^2)....\right),$$

where $K = -(-\partial^\sigma\partial_\sigma)^{-1}\partial^\mu A_\mu^{ad}$, where A^{ad} means that in $A = A^a\tau^a$, we have the matrices τ in the adjoint representation (Sect. 11.1). The application of ghosts allows to express the rules of calculation in a way similar to (fermion) quantum electrodynamics.

For the $A_0 = 0$ gauge the formula for the $\Delta(A)$ function is

$$\Delta(A)\int dU\delta(A_0^U) = 1. \tag{11.48}$$

Now

$$A_0^U = U^{-1}A_0 U + iU^{-1}\partial_0 U.$$

Hence, on the basis of Eq. (11.31) we have

$$\Delta(A)\int \delta(A_0 + i\partial_0 UU^{-1})dU = 1. \tag{11.49}$$

It follows from Eq. (11.49) by a similar argument to the one in (11.40) that $\triangle(A) = const.$

As a comment to the quantization of the electromagnetic field we can conclude the following: concerning Sect. 9.1 we can see from Eqs. (11.43)–(11.44) that for an Abelian gauge field ($f^{abc} = 0$) the Faddeev-Popov determinant for the ξ-gauge is a constant. We were allowed to omit it in the quantization in Chàp. 9. From Eq. (11.49) it follows that the choice of the $A_0 = 0$ gauge also leads to a constant Faddeev-Popov determinant. Hence, we did not have to introduce the Faddeev-Popov determinant either in the ξ gauge or in the $A_0 = 0$ gauge for electrodynamics in Chap. 9.

11.5 The Background Field Method

In the calculation of the effective action in Sects. 7.5 and 9.4 we used an external current J_μ and applied an expansion around the classical solution

$$A_\mu = A_\mu^{cl} + \sqrt{\hbar} A_\mu^q. \tag{11.50}$$

We extend this method to gauge theories (we consider Euclidean version from now on). In Abelian gauge theories the current J_μ must be conserved $\partial_\mu J_\mu = 0$ if the equation $\partial_\mu F_{\mu\nu} = J_\nu$ is to be satisfied. In non-Abelian gauge theories this consistency condition is

$$D_\mu(A^{cl})J_\mu = 0.$$

At one loop we can repeat the argument with the saddle point expansion of Sect. 7.4. At higher loops this argument does not work. One is using an external field A^{cl} for the derivation of the effective action at higher loops but one does not require that A^{cl} satisfies the classical equations of motion [2, 3, 234].

In order that A in Eq. (11.50) transforms in a covariant way we define the transformation

$$A_\mu^{cl} \rightarrow U^{-1} A_\mu^{cl} U + i U^{-1} \partial_\mu U \equiv A_\mu^{\prime cl}, \tag{11.51}$$

$$A_\mu^q \rightarrow U^{-1} A_\mu^q U \equiv A_\mu^{\prime q}. \tag{11.52}$$

When we insert the decomposition (11.50) into the classical action then we can recognize that the gauge

$$D_\mu(A^{cl})A_\mu^q = C$$

with the gauge fixing function Φ (we suppress U in Tr^U when the trace is over the adjoint representation)

$$\Phi(A) = \int dC \delta(D_\mu(A^{cl})A^q_\mu - C) \exp\left(-\frac{1}{2\xi g^2} \int dx Tr(C^2)\right) \qquad (11.53)$$

will be useful. Let us calculate the Faddeev-Popov determinant in this gauge. Note that under the infinitesimal gauge transformation $U = 1 - iu$

$$\begin{aligned} D_\mu(A'^{cl})A'^q_\mu &= D_\mu(A'^{cl})(A'_\mu - A'^{cl}_\mu) \simeq \\ D_\mu(A^{cl})A^q_\mu &+ D_\mu(A^{cl})D_\mu(A^{cl} + \sqrt{\hbar}A^q)u \end{aligned} \qquad (11.54)$$

Hence, repeating the argument from the previous section we obtain

$$\Delta(A) = \det\left(D_\mu(A^{cl})D_\mu(A^{cl} + \sqrt{\hbar}A^q)\right) \qquad (11.55)$$

With the choice (11.53) of Φ we obtain a modification of the functional integral with the term

$$\exp\left(-\frac{1}{2\xi g^2} Tr \int dx (D_\mu(A^{cl})A^q_\mu)^2\right).$$

We consider the functional integral for the generating functional

$$\begin{aligned} Z[J] &= \\ \int dA^q \exp&\left(-\frac{1}{4g^2\hbar} \int Tr F_{\mu\nu} F_{\mu\nu}(A^{cl}_\mu + \sqrt{\hbar}A^q_\mu) - \frac{1}{\hbar g^2} \int J_\mu(A^{cl}_\mu + \sqrt{\hbar}A^q_\mu)\right) \\ &\times \exp\left(-\frac{1}{2\xi g^2} Tr \int dx (D_\mu(A^{cl})A^q_\mu)^2\right) \det\left(D^\mu(A^{cl})D_\mu(A^{cl} + \sqrt{\hbar}A^q)\right). \end{aligned}$$
$$(11.56)$$

In Sect. 11.4 we have derived the functional integral for gauge invariant functionals $\mathcal{F}$. $Z[J]$ is not gauge invariant. However, it is understood in Eq. (11.56) that we use $Z[J]$ in order to calculate the expectation values of gauge invariant expressions as for example the Wilson loops $Tr U^A(x, x)$ or the effective action. Then, the discussion of Sect. 11.4 applies.

11.6 The Effective Action in Non-Abelian Gauge Theories

For a computation of the effective action (in Euclidean field theory) at one loop we follow Sect. 7.5. We choose the current $J_\mu(A^{cl})$ and the background field A^{cl} such that

$$D_\mu(A^{cl})F_{\mu\nu}(A^{cl}) = J_\nu(A^{cl}), \qquad (11.57)$$

where

$$D_\mu(A^{cl})J_\mu(A^{cl}) = 0. \qquad (11.58)$$

We write the field A as in Eq. (11.50) and expand the classical action around A_μ^{cl}. We obtain in the exponential of Eq. (11.56)

$$
\begin{aligned}
\frac{1}{4g^2\hbar}\left(\int Tr F_{\mu\nu}F_{\mu\nu}(A) + \int Tr(JA)\right) &+ \frac{1}{2\xi g^2}\int Tr(D_\mu^{cl} A_\mu^q)^2 \\
&= \frac{1}{\hbar}W^{(0)}(A^{cl}) + \frac{1}{2g^2}Tr\int A_\mu^q\left(-\delta_{\mu\nu}D_\sigma^{cl}D_\sigma^{cl} + 2i F_{\mu\nu}^{cl} + (1-\tfrac{1}{\xi})D_\mu^{cl}D_\nu^{cl}\right)A_\nu^q \\
&\equiv \frac{1}{\hbar}W^{(0)}(A^{cl}) + \frac{1}{2g^2}Tr\int A^q M A^q,
\end{aligned}
$$

(11.59)

where $D_\mu^{cl} = D_\mu(A^{cl})$ and

$$
W^{(0)}(A^{cl}) = \frac{1}{4g^2}\left(Tr\int F_{\mu\nu}^{cl} F_{\mu\nu}^{cl} + Tr\int J_\mu A_\mu^{cl}\right).
$$

The one loop formula for the generating functional in gauge theories reads (Euclidean version)

$$
Z[J] = \exp\left(-\frac{1}{\hbar}W^{(0)}(A^{cl})\right)\det\left(D_\mu^{cl}D_\mu^{cl}\right)\det(M)^{-\frac{1}{2}}.
$$

(11.60)

The effective action is

$$
\Gamma(A) = \Gamma^{(0)}(A^{cl}) - \hbar Tr\left(\ln(D_\mu^{ad}(A^{cl})D_\mu^{ad}(A^{cl}))\right) + \frac{\hbar}{2}Tr^{ad}\ln M,
$$

(11.61)

where

$$
\Gamma^{(0)}(A^{cl}) = \frac{1}{4g^2}\int Tr^{ad} F_{\mu\nu}^{cl} F_{\mu\nu}^{cl}
$$

In Eq. (11.61) we underlined that the gauge fields in the Faddeev-Popov determinant and in M in Eq. (11.59) are in the adjoint representation. If the scalar field is added than we should add to the effective action

$$
\frac{\hbar}{2}Tr\ln\left(D_\mu^{cl}D_\mu^{cl} + \frac{3}{2}\lambda^2\phi_c^+\phi_c + m^2\right),
$$

(11.62)

where the trace is the trace of the matrix differential operator which depends on the representation U of $\mathcal{G}$ acting upon the scalar field ϕ. We have studied the scalar determinant already in Sect. 11.3. There remains to calculate $\det M$. For this purpose we consider the heat kernel of M (M is a matrix of differential operators with the matrix elements $M_{\mu\nu}^{ab}$ where the lower indices come from the Euclidean vector A_μ and the upper indices from the algebra of the group $\mathcal{G}$). We have shown in [135] that the counterterms and the large field behavior of $\det M$ do not depend on ξ. We calculate the heat kernel and the determinant for $\xi = 1$ (Feynman gauge)

$$\exp\left(-\frac{\tau}{2}M\right)(x,x)_{\mu\nu} = \int DW_{(x,x)}^{\tau}(w_x)T_{\tau}^{A}T\left(\exp\left(2i\int_{0}^{\tau}ds\,F(w_x(s))\right)\right)_{\mu\nu}$$
(11.63)

where F is a matrix with matrix elements $F_{\mu\nu}$ where $F_{\mu\nu}$ is a matrix in the adjoint representation of the gauge group. The time-ordered exponential $T^{F}(\tau)$ in Eq. (11.63) is the matrix which is the solution of the equation

$$dT^{F}(\tau)_{\mu\sigma} = 2i\,F_{\mu\rho}(w_x(\tau))T_{\tau}^{F}(\tau)_{\rho\sigma}.$$

We expand the expression (11.63) for small τ. $Tr\,F = 0$ hence the term linear in τ is vanishing. We have

$$\sum_{\mu}Tr(F^2)_{\mu\mu} = Tr^{ad}\,F_{\mu\nu}F_{\nu\mu} = -Tr^{ad}\,F_{\mu\nu}F_{\mu\nu}.$$
(11.64)

We shall denote the trace over the adjoint representation and over the space index μ by Tr_{μ}^{ad}.

The τ^2-term in (11.63) has an opposite sign to the analogous term from T_{τ}^{A}. Hence, the expression (11.63) with the subtracted F^2 counterterm in the representation of the determinant (7.62) and (7.70) according to the formula of Sect. 7.6 (behaving as τ^3 for small τ and therefore leading to a finite expression for the determinant on the basis of the expansions (11.13) and (11.28)) is

$$\frac{1}{2}(2\pi)^{-2}E\left[Tr_{\mu}^{ad}\left(T_{\tau}^{A}T\left(\exp(2i\int_{0}^{\tau}ds\,F(w_x(s)))\right) - 1 - \tau^2(\frac{1}{48} - 2)F^2\right)\right],$$
(11.65)

where $\frac{1}{2}$ comes from the square root in $(\det M)^{-\frac{1}{2}}$, the trace is over the algebra of the gauge group in the adjoint representation and over the vector indices μ (note that $Tr_{\mu}1 = 4$). The corresponding contribution of the Faddeev-Popov determinant is

$$-(2\pi)^{-2}E\left[Tr^{ad}\left(T_{\tau}^{A} - 1 + \frac{\tau^2}{48}F_{\mu\nu}F_{\mu\nu}\right)\right].$$
(11.66)

The contribution of the scalar determinant to the effective action (11.61) is

$$\frac{1}{2}(2\pi)^{-2}E\left[Tr^{U}\left(T_{\tau}^{A} - 1 + \frac{\tau^2}{48}F_{\mu\nu}F_{\mu\nu}\right)\right].$$
(11.67)

The counterterms in (11.65) and (11.66) have an opposite sign to the one in the scalar determinant (11.67). The trace is over the representation U of the gauge group. Without the scalar fields the counterterm (11.65) together with an analogous term from the Faddeev-Popov determinant give a renormalization of the coupling constant of the form

$$\frac{1}{g_{bare}^2} = \frac{1}{g^2} + K\int_{\Lambda}^{1}\frac{d\tau}{\tau},$$
(11.68)

where $K > 0$ and Λ is the ultraviolet cutoff. Hence, in contradistinction to the renormalization of ϕ^4 (Eq. (3.58) in Sect. 3.6 and Eq. (7.72) in Sect. 7.6) and scalar electrodynamics of Sect. 9.4 (see Eq. (9.53) and the discussion at the end of Sect. 9.4) the bare coupling is positive and tends to zero as the cutoff $\Lambda \to \infty$ (asymptotic freedom).

The $Tr^U(F^2)$ contribution is growing with the dimension of the representation U acting upon the scalar fields (and other matter fields in the Standard Model). It is decreasing the constant K in Eq. (11.68). Hence, the asymptotic freedom fails for a large number of matter fields (when K becomes negative). Summarizing the results of this section: because of the contribution of $\ln \det M^{-\frac{1}{2}}$ and the Faddeev-Popov determinant $\ln \det D_\mu D_\mu$ (in the adjoint representation), which have an opposite sign to the contribution of the scalar field determinant, the total F^2 counterterm may have an opposite sign to the one corresponding to the Abelian Higgs model in Sect. 9.4. The positive value of K in Eq. (11.68) is called the asymptotic freedom of non-Abelian gauge theories and is of profound importance for the stability of quantum gauge theories [120, 197]. It has as a consequence the behavior of $\Gamma(A)$ in Eq. (11.61) as $K F^2 \ln F^2$ for large F with $K > 0$ [135]. The asymptotic freedom is also important for quark confinement in the continuum limit of lattice gauge theories as we may expect that we can let the lattice spacing $\delta \to 0$ (entering the gauge action as $\frac{1}{g^2(\delta)}$) while preserving the confinement present at large coupling $g(\delta)$ on the lattice (Chap. 12).

11.7 Exercises

11.1 Calculate the term bilinear in $\tilde\phi$ and in A_μ in the Lagrangian (11.10) after the shift $\phi = \tilde\phi + v$ where v is the minimum of ϕ in the Lagrangian (11.10).
11.2 Prove Eq. (11.17) for the heat kernel which is

$$d \exp\left(\frac{t}{2}\Delta_A\right)\psi = \frac{1}{2}\Delta_A \exp\left(\frac{t}{2}\Delta_A\right)\psi dt$$

repeating the derivation of Eq. (5.65).
11.3 Compute $\det(-\Delta_A + m^2)$ in a formal perturbative expansion in A_μ (as suggested for the scalar field in Sect. 7.6). Compare with the heat kernel calculus of Sect. 9.4.
11.4 Prove that there is no photon mass renormalization (Eq. (11.22) gives zero)
11.5 Discuss the small τ behavior in Eqs. (11.28) and (11.65).
11.6 Calculate the constant K in Eq. (11.67).
11.7 Prove that the derivative $D_\mu F_{\sigma\rho}$ is covariant under gauge transformation.

11.8 Show that if $\partial^\mu A_\mu = 0$ and $\partial^\mu A_\mu^U = 0$ then the Faddeev-Popov operator $\partial^\mu D_\mu(A)$ has zero modes u (i.e., $\partial^\mu D_\mu(A)u = 0$), hence it is not invertible.

11.9 Calculate the quadratic part in A_μ^q of the gauge field Lagrangian (11.59).

11.10 Prove the formula (11.63) for the heat kernel by means of differentiation over τ.

Chapter 12
Lattice Approximation

Abstract The lattice approximation gives a discrete finite dimensional approximation to Euclidean field theories. It is a useful ultraviolet cutoff. In contradistinction to other ultraviolet regularizations in Euclidean field theories it breaks the Euclidean invariance. The virtue of the lattice approximation is that it preserves the Osterwalder-Schrader (OS) positivity. If the Euclidean invariance is restored in the limit when the lattice spacing tends to zero then the main requirements for the continuation to QFT in Minkowski space will be satisfied. We define the lattice approximation for scalar fields in d dimensions. A relation to the continuous spin Ising model of statistical mechanics is explained. We show the OS positivity. We represent the lattice field by the continuum free field so that the continuum limit can be proved in perturbation theory. We develop the heat kernel representation on the lattice in terms of the compound Poisson process obtaining the polymer representation of the Higgs model discussed earlier in the continuum. We discuss the lattice approximation of gauge theories as a way to make the gauge field quantization a mathematically precise theory. We discuss the strong coupling expansion leading to the Higgs mechanism and to the fulfillment of the Wilson criterion for quark confinement.

The lattice approximation in Euclidean field theories has been introduced and investigated for scalar field theories in [124] and in gauge theories by Wilson [245]. It is a useful ultraviolet cutoff. The lattice has still another benefit important from the point of view of numerical methods: it delivers a finite dimensional approximation to infinite dimensional integrals. In contradistinction to other ultraviolet regularizations in Euclidean field theories it breaks the Euclidean invariance. The virtue of the lattice approximation is that it is OS positive. If the Euclidean invariance is restored in the limit when the lattice spacing tends to zero then the main requirements for the continuation to QFT in Minkowski space will be satisfied. In Sect. 12.1 we define the lattice approximation for scalar fields in d dimensions. A relation to the continuous spin Ising model of statistical mechanics is explained. We show the OS positivity. We represent the lattice field by a continuum field so that the continuum limit can be proved in perturbation theory. In Sect. 12.2 we develop the Feynman-Kac formula on the lattice in terms of the compound Poisson process obtaining the polymer representation of the Higgs model discussed earlier in the continuum. We discuss the perturbative continuum limit of scalar field theories. In Sect. 12.3 we consider the lattice approximation of gauge theories explaining how the Faddee-Popov procedure can be made mathematically precise. We discuss the behavior at large and small coupling.

12.1 Lattice Approximation in Euclidean Scalar Field Theory

We formulate the lattice approximation in d dimensions as such theories lead to interesting models of statistical physics with the relevant dimension $1 \leq d \leq 3$. We divide the Euclidean space-time R^d into boxes of volume δ^d. The scalar field ϕ depends on vertices of these boxes (called sites). So we write $\phi(\mathbf{x})$ as $\phi(n_1 \delta, \ldots, n_d \delta)$, where n_j are integers (we write $\mathbf{n} \in Z^d$).

For a scalar field (in Euclidean space) we replace the derivative in the Lagrangian by the lattice derivative

$$(\nabla_j^\delta \phi)(\mathbf{n}\delta) = \delta^{-1}\Big(\phi(\ldots, n_j\delta + \delta, \ldots) - \phi(\ldots, n_j\delta, \ldots)\Big). \qquad (12.1)$$

The lattice Laplacian can be written as

$$(\triangle_\delta \phi)(\mathbf{n}\delta) = \delta^{-2}\Big(\sum_j \phi(\mathbf{n}\delta + \mathbf{e}_j\delta) - 2d\phi(\mathbf{n}\delta)\Big), \qquad (12.2)$$

where $\mathbf{e}_j$ is the unit vector in the jth direction and the sum is over the nearest neighbors of the point $\mathbf{n}\delta$ in all directions $\mathbf{e}_j$. The free Euclidean action is (we denote the action and the Lagrangian by the same letter as on the lattice the action is a sum of Lagrangians)

$$\mathcal{L}_0 = -\frac{1}{2}\delta^d \sum_{\mathbf{n}} \phi(\mathbf{n}\delta)(\triangle_\delta \phi)(\mathbf{n}\delta) + \frac{m^2}{2}\delta^d \sum_{\mathbf{n}} \phi(\mathbf{n}\delta)^2. \qquad (12.3)$$

The interaction has the form

$$\mathcal{L}_I = \sum_{\mathbf{n}} U(\phi(\mathbf{n}\delta))\delta^d. \qquad (12.4)$$

The quantum (Euclidean) field theory is defined as usual by the integral ($\mathcal{L} = \mathcal{L}_0 + \mathcal{L}_I$)

$$d\mu_\delta = \prod_{\mathbf{n}} d\phi(\mathbf{n}\delta) \exp(-\mathcal{L}). \qquad (12.5)$$

It is useful to treat a function of $\mathbf{n} \in Z^d$ as a coefficient of the Fourier series of a periodic function of the momentum with a period $\frac{2\pi}{\delta}$

$$\tilde{\phi}_\delta(\mathbf{k}) = (2\pi)^{-\frac{d}{2}} \sum_\mathbf{n} \phi(\mathbf{n}\delta) \exp(i\mathbf{k}\mathbf{n}\delta)$$

Then, $\phi(\mathbf{n}\delta)$ can be expressed by $\tilde{\phi}_\delta(\mathbf{k})$ as

$$\phi(\mathbf{n}\delta) = (2\pi)^{-\frac{d}{2}} \int\limits_{-\frac{\pi}{\delta}}^{\frac{\pi}{\delta}} d\mathbf{k} \exp(i\mathbf{k}\mathbf{n}\delta)\tilde{\phi}_\delta(\mathbf{k}),$$

where the $\mathbf{k}$-integration is over the cube $-\frac{\pi}{\delta} \le k_j \le \frac{\pi}{\delta}$.

If the lattice is infinite then the measure $\prod_\mathbf{n} d\phi(\mathbf{n}\delta) \exp(-\mathcal{L}_0)$ should be understood as the Gaussian measure on an infinite dimensional space of sequences $\phi(\mathbf{n}\delta)$ as discussed in Sect. 1.3. We can calculate from Eq. (12.5) in free field theory in an infinite volume (where the product $\prod_\mathbf{n} d\phi(\mathbf{n}\delta)$ is over all sites of the lattice)

$$\begin{aligned}
< \phi(\mathbf{n}\delta)\phi(\mathbf{n}'\delta) > &= Z^{-1} \int \prod_\mathbf{n} d\phi(\mathbf{n}\delta) \exp(-\mathcal{L}_0)\phi(\mathbf{n}\delta)\phi(\mathbf{n}'\delta) \\
&= (2\pi)^{-d} \int_{-\frac{\pi}{\delta}}^{\frac{\pi}{\delta}} d\mathbf{k} \exp(i(\mathbf{k}\mathbf{n} - \mathbf{k}\mathbf{n}')\delta)\omega_\delta^{-2},
\end{aligned} \quad (12.6)$$

where

$$Z = \int \prod_\mathbf{n} d\phi(\mathbf{n}\delta) \exp(-\mathcal{L}_0)$$

and

$$\omega_\delta^2(\mathbf{k}) = \left(2d - 2\sum_{j=1}^{d} \cos(\delta k_j)\right)\delta^{-2} + m^2. \quad (12.7)$$

It is understood that in Eq. (12.6) we first consider a finite volume cutoff and subsequently take the infinite volume limit (as explained in Sect. 1.3 in a discussion of Gaussian measures in an infinite number of dimensions). Equation (12.6) follows because in Fourier transforms the free Lagrangian is just

$$\mathcal{L}_0 = \frac{1}{2} \int\limits_{-\frac{\pi}{\delta}}^{-\frac{\pi}{\delta}} d\mathbf{k}\tilde{\phi}_\delta(\mathbf{k})\tilde{\phi}_\delta(-\mathbf{k})\omega_\delta^2.$$

The restriction of the momentum to the interval $[-\frac{\pi}{\delta}, \frac{\pi}{\delta}]$ in Eq. (12.6) is a consequence of the periodicity with respect to lattice translation $\mathbf{n}\delta$ (where $\mathbf{n}$ is a d-tuple of integers) of the lattice Lagrangian and as a consequence ω_δ is invariant under the translation $k_j \to k_j + \frac{2\pi}{\delta}$. We can express the lattice field by the continuum field. Let $\tilde{\phi}(\mathbf{k})$ be the Fourier transform of the continuum Euclidean field $\phi(\mathbf{x})$, $\mathbf{x} \in R^d$. Then, we define

$$\phi(\mathbf{n}\delta) = (2\pi)^{-\frac{d}{2}} \int_{-\frac{\pi}{\delta}}^{\frac{\pi}{\delta}} d\mathbf{k} \exp(i\mathbf{k}\mathbf{n}\delta) \omega_\delta^{-1}(\mathbf{k}) \omega(\mathbf{k}) \tilde{\phi}(\mathbf{k}), \qquad (12.8)$$

where

$$\omega(\mathbf{k})^2 = m^2 + \mathbf{k}^2.$$

We can check that in fact the covariance of the Gaussian field (12.8) coincides with (12.6)

$$\int d\mu_0(\phi)\phi(\mathbf{n}\delta)\phi(\mathbf{n}'\delta) = (2\pi)^{-d} \int_{-\frac{\pi}{\delta}}^{\frac{\pi}{\delta}} d\mathbf{k} \exp(i(\mathbf{k}\mathbf{n} - \mathbf{k}\mathbf{n}')\delta)\omega_\delta^{-2}, \qquad (12.9)$$

where μ_0 is the Gaussian measure defined in Eqs. (3.66)–(3.68) for the continuum free scalar field (it has the covariance $\delta(\mathbf{k} + \mathbf{k}')\omega(k)^{-2}$). From Eq. (12.8) it follows that the lattice can be considered as a particular ultraviolet cutoff (because the range of $\mathbf{k}$ is finite). In such a case the continuum propagator is replaced by the lattice propagator

$$\left(\mathbf{k}^2 + m^2\right)^{-1} \rightarrow \left(\left(2d - 2\sum_{j=1}^{d} \cos(\delta k_j)\right)\delta^{-2} + m^2\right)^{-1}.$$

In a renormalizable quantum field theory (e.g., ϕ^4 in four dimensions) introducing the normal ordering (as in Eq. (3.14))

$$: \exp(\lambda\phi(\mathbf{n}\delta) := \exp(\lambda\phi(\mathbf{n}\delta) < \exp(\lambda\phi(\mathbf{n}\delta) >^{-1}$$

and proper counterterms for the mass and charge renormalization we can prove the continuum limit in each order of the perturbation expansion [25, 210]. In detail,

$$\int d\mu_0(\phi)\phi(\mathbf{n}\delta)...\phi(\mathbf{n}'\delta) \sum_{r=0}^{\infty} \frac{1}{r!}\left(-\sum U\delta^d\right)^r \rightarrow S(\mathbf{x}, ..., \mathbf{x}'), \qquad (12.10)$$

where on the lhs we have a sum $\sum U$ over a finite volume and the rhs is understood as the perturbation formula for the continuum Schwinger functions (as discussed in Sect. 3.7). When we represent the lattice fields by continuum fields (as in Eq. (12.8)) then the proof of Eq. (12.10) goes in the same way as in continuum ϕ^4 (in particular in $d = 2$ the normal ordering is sufficient for the proof of the limit $\delta \rightarrow 0$).

As can be seen from Eq. (12.3) the lattice approximation has the form of a continuous spin Ising model with a ferromagnetic nearest neighbor coupling. This allows a study of the field theoretic model by means of methods of the statistical mechanics [124, 221]. The immediate benefit of using lattice approximation is that the continuum limit satisfies the OS positivity. So, if we can also prove that the continuum limit is Euclidean invariant then the main conditions for an analytic continuation to a relativistic quantum field theory will be satisfied.

The OS positivity of lattice theory follows from the argument outlined heuristically in Sect. 3.10. We divide the space into two half-spaces: E_+ with $n_1 > 0$, E_- with $n_1 < 0$ and the separating hyperplane E_0 with $n_1 = 0$ (the first axis plays the role of the time axis). We define $\theta\phi(n_1\delta, , , , , , , n_d\delta) = \phi(-n_1\delta, , , , , , , n_d\delta)$. Then

$$\sum_{\mathbf{n}} \mathcal{L}(\mathbf{n}\delta)) = \sum_{E_+} \mathcal{L}(\mathbf{n}\delta)) + \theta \sum_{E_+} \mathcal{L}(\mathbf{n}\delta) + \sum_{E_0} \mathcal{L}_{E_0}(\mathbf{n}\delta) + \sum \phi\theta\phi$$
$$\equiv \mathcal{L}_+ + \theta\mathcal{L}_+ + \sum_{E_0} \mathcal{L}_{E_0} + \sum \phi\theta\phi, \qquad (12.11)$$

where the last term comes from the coupling of the nearest neighbors across the $x_1 = 0$ line and $\sum_{E_0}$ denotes the sum over terms lying on the line $x_1 = 0$. Assume $f(n_1\delta, , , , , , , n_d\delta) = 0$ on E_-. We consider the generating functional

$$Z[f] = \int d\mu_\delta \exp(i\phi(f)),$$

where f is a real function on E_+ and $\phi(f) = \sum_{\mathbf{n}} \phi(n_1\delta, , , , , , , n_d\delta)$ $f(n_1\delta, , , , , , , n_d\delta)$.

Neglecting the last term in Eq. (12.11) for a moment it follows that

$$\sum_{j,k} c_j c_k^* Z[f_j - \theta f_k]$$
$$= \int \prod d\phi(\mathbf{n}\delta) \exp(-\mathcal{L}) \sum_j c_j \exp(i\phi(f_j)) \sum_k c_k^* \exp(-i\phi(\theta f_k))$$
$$= \int \prod_+ d\phi(\mathbf{n}\delta) \exp(-\mathcal{L}_+ - \tfrac{1}{2} \sum_{E_0} \mathcal{L}) \sum_j c_j \exp(i\phi(f_j)) \qquad (12.12)$$
$$\times \theta \prod_+ d\phi(\mathbf{n}\delta) \Big(\exp(-\mathcal{L}_+ - \tfrac{1}{2} \sum_{E_0} \mathcal{L}_{E_0}) \sum_k c_k \exp(i\phi(f_k)) \Big)^* \geq 0$$

as the term $\mathcal{L}_{E_0}$ is invariant under reflection and the expression (12.12) is of the form $F F^*$ because the integrals over E_+ and E_- are the same.

Finally, the omitted term

$$\exp \phi\theta\phi = \sum_l \frac{1}{l!} \phi^l \theta\phi^l$$

modifies Eq. (12.12) to the form

$$\sum_{j,k} c_j c_k^* Z[f_j - \theta f_k] = \sum_l (F_l F)^* F_l F \geq 0,$$

where the form of F_l follows from Eqs. (12.11)–(12.12) with $F_l \simeq \phi^l$. As a consequence, the modification of the expression (12.12) by a sum of such terms does not break the positivity property (12.12). In the proof of the OS positivity the nearest neighbor Ising spin interaction seems essential. Nevertheless, the OS-positivity holds true [133, 149] for local perturbations of generalized free fields (whose free propagator has a mass spectrum). On the other hand a heuristic argument at the end of Sect. 3.10 which seems to apply to any local Lagrangian in fact fails for Lagrangians with higher order derivatives what could have been seen in their lattice approximations.

12.2   Polymer Representation

We return to the polymer representation of Sect. 9.3 in order to derive its lattice version. For this purpose we need the Feynman-Kac formula on the lattice. First, we must define an analog of the Brownian motion on the lattice. It is to be a stochastic process which is increasing by jumps from one site of the lattice to another site. There is such a process called the compound Poisson process ([226], this is the random walk with a probability of a jump of length n by the walker in time t given by the Poisson distribution) with the generating functional

$$E[\exp(i\mathbf{u}\mathbf{w}_t)] = \int dP^\delta(\mathbf{w})\exp(i\mathbf{u}\mathbf{w}_t) = \exp\left(t d\delta^{-2}\int d\nu(\mathbf{e})(\exp(i\mathbf{e}\mathbf{u}\delta) - 1)\right)$$
$$= \exp\left(t\delta^{-2}\left(\sum_{j=1}^d(\cos(\delta u_j) - 1)\right)\right),$$

$$(12.13)$$

where $\nu(\mathbf{e}) = \nu(-\mathbf{e})$ is a discrete probability measure on unit jumps giving them an equal probability $\frac{1}{2d}$. Let us note that the characteristic function (12.13) of the process is invariant under translations $u_j \to u_j + \frac{2\pi}{\delta}$. This is so because $\mathbf{w}_t$ takes integer values $\mathbf{w}_t = \delta(n_1(t), \dots, n_d(t))$ where $\mathbf{n} \in Z^d$ is a d-tuple of integers. A time homogeneous Markov process $\mathbf{w}_t$ defines a semigroup ($\mathbf{x} = \mathbf{n}\delta, \mathbf{n} \in Z^d$)

$$(T_t f)(\mathbf{x}) = E[f(\mathbf{x} + \mathbf{w}_t)]. \qquad (12.14)$$

As in Sect. 12.1 we treat a function on Z^d as a coefficient of an expansion of a periodic function of the momentum. If we use the momentum space representation of f and the expectation value of the characteristic function (12.13) of $\mathbf{w}_t$ then we can rewrite Eq. (12.14) as

$$E[f(\mathbf{x} + \mathbf{w}_t)] = (2\pi)^{-\frac{d}{2}} E[\int_{-\frac{\pi}{\delta}}^{\frac{\pi}{\delta}} d\mathbf{p}\exp(i\mathbf{p}\mathbf{x})\exp(i\mathbf{p}\mathbf{w}_t)]f(\mathbf{p})$$

$$(12.15)$$

$$= (2\pi)^{-\frac{d}{2}} \int_{-\frac{\pi}{\delta}}^{\frac{\pi}{\delta}} d\mathbf{p}\exp(i\mathbf{p}\mathbf{x})\exp\left(t\delta^{-2}\sum_{j=1}^d(\cos(\delta p_j) - 1)\right)f(\mathbf{p}).$$

Differentiating Eq. (12.15) we obtain

$$\frac{dT_t f}{dt}(\mathbf{n}\delta) = (2\pi)^{-\frac{d}{2}} \int_{-\frac{\pi}{\delta}}^{\frac{\pi}{\delta}} d\mathbf{p} \exp(i\mathbf{p}\mathbf{n}\delta)$$

$$\times \exp\left(t\delta^{-2} \sum_{j=1}^{d}(\cos(\delta p_j)-1)\right) f(\mathbf{p})\left(\delta^{-2} \sum_{j=1}^{d}(\cos(\delta p_j)-1)\right) \tag{12.16}$$

$$= \tfrac{1}{2}(\Delta_\delta T_t f)(\mathbf{n}\delta),$$

where Δ_δ is defined in Eq. (12.2). We can define the measure $dP_{(\mathbf{x},\mathbf{y})}^t(\mathbf{w}) = dP^\delta(\mathbf{w})\delta_\delta(\mathbf{w}_t + \mathbf{x} - \mathbf{y})$, where δ_δ is the Kronecker δ-function on Z^d. Inserting the Fourier representation of the δ-function and using (12.13) we obtain

$$K_t^\delta(\mathbf{x}-\mathbf{y}) = \int dP_{(\mathbf{x},\mathbf{y})}^t(\mathbf{w}) = \int dP^\delta(\mathbf{w})(2\pi)^{-d} \int_{-\frac{\pi}{\delta}}^{\frac{\pi}{\delta}} d\mathbf{u} \exp(i\mathbf{u}(\mathbf{w}_t + \mathbf{x} - \mathbf{y}))$$

$$= (2\pi)^{-d} \int_{-\frac{\pi}{\delta}}^{\frac{\pi}{\delta}} d\mathbf{u} \exp(i\mathbf{u}(\mathbf{x}-\mathbf{y})) \exp\left(t\delta^{-2} \sum_{j=1}^{d}(\cos(\delta u_j)-1)\right). \tag{12.17}$$

Using the definition of the kernel (Sect. 3.4 when $R^d \to Z^d$)

$$(T_t f)(\mathbf{x}) = \sum_{\mathbf{y}} K_t^\delta(\mathbf{x}, \mathbf{y}) f(\mathbf{y})$$

and comparing with Eq. (12.16) we can see that

$$K_t^\delta(\mathbf{x}-\mathbf{x}') = \exp\left(\frac{t}{2}\Delta_\delta\right)(\mathbf{x}, \mathbf{x}'), \tag{12.18}$$

where $\mathbf{x} = \delta(n_1, ..., n_d)$. $K_t^\delta(\mathbf{x}-\mathbf{x}')$ is a positively definite matrix satisfying the composition law

$$\sum_{\mathbf{y}} K_t^\delta(\mathbf{x}-\mathbf{y}) K_s^\delta(\mathbf{y}-\mathbf{z}) = K_{t+s}^\delta(\mathbf{x}-\mathbf{z}).$$

Such a matrix is a transition function of a Markov process. Using the transition function we could define a measure P^δ of Eq. (12.13) by means of an analog of the formula (5.32)

$$\int dP^\delta(w) F(\mathbf{w}(s_1), \mathbf{w}(s_2), ..., \mathbf{w}(s_n)) = \sum_{\mathbf{x}_1} .... \sum_{\mathbf{x}_n} F(\mathbf{x}_1, ..., \mathbf{x}_n)$$

$$\times K_{s_1}^\delta(\mathbf{x}_1) K_{s_2-s_1}^\delta(\mathbf{x}_2-\mathbf{x}_1)....K_{s_n-s_{n-1}}^\delta(\mathbf{x}_n-\mathbf{x}_{n-1}). \tag{12.19}$$

It follows from Eqs. (12.13)–(12.19) that in the limit $\delta \to 0$ we get $\int dP_{(\mathbf{x},\mathbf{y})}^t \to (2\pi t)^{-\frac{d}{2}} \exp(-(\frac{\mathbf{x}-\mathbf{y})^2}{2t}))$, the compound Poisson process tends to the Brownian motion and $dP_{(\mathbf{x},\mathbf{y})}^t \to dW_{(\mathbf{x},\mathbf{y})}^t$. We consider the Feynman-Kac formula for the operator

$$H_\delta = -\frac{1}{2}\Delta_\delta + V(\mathbf{n}\delta). \tag{12.20}$$

We obtain

$$(\exp(-tH_\delta)\psi)(\mathbf{n}\delta) = \int dP^\delta(\mathbf{w}) \exp\Big(-\int_0^t V(\mathbf{n}\delta + \mathbf{w}_s)ds\Big)\psi(\mathbf{n}\delta + \mathbf{w}_t).$$

$$(12.21)$$

The rhs of Eq. (12.21) defines a semigroup. In order to prove Eq. (12.21) it is sufficient to calculate the generator of this semigroup. This requires just a calculation of the time derivative of Eq. (12.21) at $t = 0$ using Eq. (12.16). Equation (12.21) could also be proved as a version of the Trotter product formula of Sect. 5.1.

The evolution kernel is expressed as

$$(\exp(-tH_\delta)\psi)(\mathbf{x}, \mathbf{y}) = \int dP^t_{(\mathbf{x},\mathbf{y})}(\mathbf{w}) \exp\Big(-\int_0^t V(\mathbf{w}_s)ds\Big). \qquad (12.22)$$

We can now repeat the polymer representation of Symanzik [229] on the lattice (for a real scalar field)

$$\exp\Big(-\frac{\lambda^2}{8}\delta^d \phi(\mathbf{n}\delta)^4\Big) = \int d\Phi(\mathbf{n}\delta) \exp\Big(-\frac{1}{2}\delta^d \Phi(\mathbf{n}\delta)^2\Big) \exp\Big(\frac{i\lambda}{2}\delta^d \phi(\mathbf{n}\delta)^2\Phi(\mathbf{n}\delta)\Big)$$

$$(12.23)$$

As in Eq. (9.27) we obtain the representation of the two-point function in ϕ^4 model on the lattice (we skip the electromagnetic field in this section we return to it in the next section)

$$< \phi(\mathbf{x})\phi(\mathbf{y}) >= \int \prod_\mathbf{n} d\Phi(\mathbf{n}\delta) \exp(-\tfrac{1}{2}\delta^d \Phi(\mathbf{n}\delta)^2)$$
$$\int_0^\infty dt \exp(-\tfrac{1}{2}m^2 t)dP^t_{(\mathbf{x},\mathbf{y})}(\mathbf{w}) \exp(-i\lambda \int_0^t \Phi(\mathbf{w}_s)ds) \det M^{-\frac{1}{2}}, \qquad (12.24)$$

where

$$\det M = \exp\Big(-\sum_\mathbf{x} \int_0^\infty \frac{dt}{t} dP^t_{(\mathbf{x},\mathbf{x})}(w)\Big(\exp(-i\lambda \int_0^t \Phi(\mathbf{w}_s)ds) - 1\Big)\Big). \qquad (12.25)$$

Note that $K^\delta_t(\mathbf{x} - \mathbf{y})$ is a bounded function of t. Then, there are no divergencies at small t in Eq. (12.25). When we expand the exponential of the determinant (12.25) then we can calculate the Gaussian integral over Φ. Φ is the Gaussian variable with the covariance

$$< \Phi(\mathbf{n}\delta)\Phi(\mathbf{n}'\delta) >= \delta^{-d}\delta(\mathbf{n}, \mathbf{n}'),$$

where $\delta(\mathbf{n}, \mathbf{n}')$ is the Kronecker δ. In the zeroth order of the expansion of the determinant we obtain in Eq. (12.24)

$$< \phi(\mathbf{x})\phi(\mathbf{y} >\simeq \int_0^\infty dt \exp(-\tfrac{1}{2}m^2 t)dP^t_{(\mathbf{x},\mathbf{y})}(\mathbf{w})$$
$$\times \exp\Big(-\tfrac{\lambda^2}{2}\int_0^t \int_0^t dsds' \delta^{-d}\delta(w_s, w_{s'})\Big)$$

The result is similar to that of Sect. 9.3 (Eq. (9.28)) but instead of a singular Dirac δ function we obtain the discrete (Kronecker) δ_δ. The polymer representation has been applied in [44, 45] in order to prove the mass gap in ϕ^4 in two dimensions as well as to show the triviality of ϕ^4 model in more than four dimensions [95].

12.3 Lattice Approximation in Gauge Theories

For a formulation of lattice gauge theories we need variables which transform in a covariant way under the gauge transformations. If we denote the group element $g_{yx} = U^A(y, x)$ constructed in Sect. 11.1 as a line integral (11.7) from x to y then under the gauge transformation ($g_x \in \mathcal{G}$)

$$g_{yx} \rightarrow g_y^{-1} g_{yx} g_x \tag{12.26}$$

and $g_{xy} = g_{yx}^{-1}$. These are the field variables for the gauge theory on the lattice. We integrate with the Haar measure $\prod_{<x,y>} dg_{xy}$ in the Euclidean path integral. As in the case of scalar fields in Sect. 12.1 we divide the Euclidean space into boxes. Then, we consider a directed line from the point x to y on the lattice (of length δ, called a bond and denoted as $< x, y >$). We associate to it a group element g_{yx}. We need to define an action which will be invariant under the gauge transformations. It can be seen from Eq. (11.9) that a trace of the integral (11.7) along a closed curve is gauge invariant. The shortest closed curve on the lattice comes from a boundary of the square P (called plaquette) formed by elementary bonds on the boundary of the box as illustrated in (12.27). The trace of the plaquette $g_{xy} g_{yu} g_{uv} g_{vx}$ could be a candidate for the lattice action but it depends on the group representation

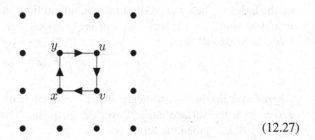

$$\tag{12.27}$$

(in the heuristic limit to the continuum there is no such a dependence). We can define the lattice in a representation independent way as

$$-\mathcal{L} = \frac{1}{g^2} \sum_P \chi(g_{xy} g_{yu} g_{uv} g_{vx}) + (\chi(g_{xy} g_{yu} g_{uv} g_{vx}))^*, \tag{12.28}$$

where g is a coupling constant, χ is the character on the group, i.e., a sum of traces of irreducible group representations U, $\chi(g) = \sum_U Tr(U(g))$. The action (12.28) is not unique. There are many expressions for the lattice action which in a formal continuum limit give the same action (11.4). Nevertheless, the various actions may have different physical consequences [218].

The gauge invariant coupling to the scalar field ϕ transforming according to the representation U ($\phi(y) \rightarrow (U(g(y)))^{-1}\phi(y)$) of the gauge group is defined as

$$\sum_{<x,y>} \phi^+(x)U(g_{xy})\phi(y), \tag{12.29}$$

where the sum is over all bonds $< x, y >$.

The Higgs interaction is a sum over sites x

$$\sum_x \left(\frac{1}{8}\lambda^2(\phi^+\phi)^2 - \frac{\mu^2}{2}\phi^+\phi \right). \tag{12.30}$$

In the Abelian case $\mathcal{G} = U(1)$ we have $g_{xy} = \exp(iA_{xy})$. The action (12.28) is

$$\mathcal{L} = \frac{1}{g^2}\sum_P \left(1 - \cos(A_{xy} + A_{yu} + A_{uv} + A_{vx}) \right) \tag{12.31}$$

The integral in the path integral is over $A_{xy} \in [-\pi, \pi]$. The gauge transformation in the Abelian theory reads

$$A'_{xy} = A_{xy} + f_y - f_x. \tag{12.32}$$

We could expand the cosine in Eq. (12.31) in a power series taking only the quadratic term for the action (this is called the non-compact form of the Abelian gauge theory on the lattice). Then, the proof of the continuum limit in the perturbative expansion would be similar to the scalar case of the previous section. In the Abelian case we have the Stokes theorem

$$\int A_\mu dx^\mu = \int F_{\mu\nu}d\sigma^{\mu\nu}, \tag{12.33}$$

where on the lhs there is an integral over a closed curve on the boundary of a plaquette and $d\sigma^{\mu\nu}$ is the surface integral over the plaquette. Then, in Eq. (12.31), if $g_{yx} = \exp(i\int_x^y A_\mu dx^\mu)$ then the action is

$$\exp\left(i\int F_{\mu\nu}d\sigma^{\mu\nu} \right) + \exp\left(-i\int F_{\mu\nu}d\sigma^{\mu\nu} \right) \simeq -\left(\int F_{\mu\nu}d\sigma^{\mu\nu} \right)^2$$

Representing the lattice field by the continuum electromagnetic field we recover in Abelian gauge theories the continuum limit with non-compact form of the action. However, there are non-perturbative phenomena associated with the compact form of the action [198, 218].

In the Abelian Higgs model (12.29) we can write the heat kernel formula for the lattice Laplacian coupled to the Abelian gauge field

$$\exp(\tfrac{t}{2}\Delta_\delta^A)(\mathbf{x}, \mathbf{y}) = \int dP_{(\mathbf{x},\mathbf{y})}^t(\mathbf{w}) \exp\left(i \int_0^t A_{\mathbf{w}_{s+ds}\mathbf{w}_s}(\mathbf{w}_{s+ds} - \mathbf{w}_s)\right) \qquad (12.34)$$

where $\mathbf{w}$ is the compound Poisson process on the lattice discussed in Sect. 12.2. In Eq. (12.34) the integral with $\mathbf{w}_s$ is changing between the neighboring sites. The expectation values in Abelian and non-Abelian Higgs model are calculated in [74] using the random walk representation. The representation (12.34) is inconvenient for calculations. The result could be obtained in another way expanding the heat kernel $\exp(\tfrac{t}{2}\Delta_A)$ in polynomials of group elements. We have

$$(\Delta_A f)(\mathbf{n}\delta) = (\Delta f)(\mathbf{n}\delta) + \sum_{\mathbf{n}'} \exp(i A_{\mathbf{n}\mathbf{n}'}) f(\mathbf{n}'\delta), \qquad (12.35)$$

where the sum is over the nearest neighbors. Using the Dyson expansion we obtain

$$\begin{aligned}\exp(\tfrac{t}{2}\Delta_\delta^A) &= \exp(\tfrac{t}{2}\Delta_\delta) + \int_0^t ds \exp(\tfrac{t-s}{2}\Delta_\delta) \exp(i A) \exp(\tfrac{s}{2}\Delta_\delta) \\ &+ \int_0^t ds \exp(\tfrac{t-s}{2}\Delta_\delta) \exp(i A) \int_0^s ds' \exp(\tfrac{s-s'}{2}\Delta_\delta) \exp(i A) \exp(\tfrac{s'}{2}\Delta_\delta) + \cdots\end{aligned}$$
$$(12.36)$$

The action of $\exp(i A)$ in Eq. (12.36) is defined in Eq. (12.35). Eq. (12.36) gives a meaning to the stochastic integral in Eq. (12.34). However, the formula (12.34) has also a non-perturbative definition. Using the heat kernel representation we can obtain the formula for the one-loop scalar determinant in lattice gauge theories analogous to Eq. (9.33) (for another representation of the lattice determinant see [219])

The main question of Euclidean lattice field theories remains: whether the OS positivity is satisfied and whether an Euclidean invariant continuum limit exists. Then, on the basis of the Osterwalder-Schrader reconstruction theorem [191] we may expect that such a theory has an analytic continuation to the Minkowski space satisfying Wightman axioms [228] (with a proper modification suitable for gauge theories [218]). A useful technical trick for the proof of reflection positivity [192, 218] in lattice gauge theory is to set the gauge fields on the lines (between bonds)

$$x_j = \left(n_j + \frac{1}{2}\right)\delta.$$

Another technical tool is the temporal gauge setting all g_{xy} along the x^1 line (x^1 is the Euclidean time) to 1 [192]. That this is possible can easily be seen in the Abelian case as from Eq. (12.32) we can choose (starting from a particular initial site x)

$$f_y = f_x - A_{xy}.$$

when $< x, y >$ lie parallel to the x^1 axis (in fact we can make $A' = 0$ along any non-intersecting curve on the lattice). After the choice of the temporal gauge the proof of the OS positivity is similar to the one for the scalar field in Sect. 12.1 [192].

The lattice gauge theories at the strong coupling have some properties required in QCD and in the Standard Model at strong coupling (confinement of quarks and the Higgs mechanism of mass generation). As shown in the paper by Wilson [245] (introducing the lattice approximation) the confinement can be associated with the behavior of the expectation value of the Wilson loop variable (discussed in Sect. 11.1) $L(\partial S) = < Tr(U^A(x, x)) >$, where the path-ordered integral is over a closed curve ∂S (comprising a surface S). If $L(\xi)$ is decaying as an area $|S|$ of the minimal surface S with the boundary $\xi = \partial S$ then the force between particles interacting with the gauge field is rising linearly with the distance. Wilson has shown that at strong coupling even in Abelian gauge theory on the lattice there is the area decay of $L(\partial S)$. The argument is based on the simple integral

$$\int_{-\pi}^{\pi} \exp(inx)dx = 2\pi \delta_{n0}. \tag{12.37}$$

Let us consider for simplicity a flat lattice loop ξ formed from bond integrals (of length δ) and the closed curve integral $\int A_\mu d\xi_\mu$ represented as $A_{x_1 x_2} + \cdots + A_{x_k x_1}$. If we expand the action (12.31) in powers of g^{-2} then owing to the condition (12.37) the lowest order non-vanishing term arises when

$$A_{x_1 x_2} + \cdots + A_{x_k x_1} \pm \sum_P (A_{xy} + A_{yu} + A_{uv} + A_{vx}) = 0. \tag{12.38}$$

It can easily be seen that the minimal number of plaquettes required for the fulfillment of the condition (12.38) is equal to $\delta^{-2}|S|$. Hence

$$< \exp\left(i(A_{x_1 x_2} + \cdots + A_{x_k x_1})\right) > \simeq (g^{-2})^{\delta^{-2}|S|}. \tag{12.39}$$

The convergence of the strong coupling expansion has been proved in Abelian as well as non-Abelian gauge theories [192, 218]. There are also results on the lattice supporting the Higgs mass generation mechanism at the strong coupling [192]. There remains the question what happens when g^2 is decreasing as required for the continuum limit. We expect that the area law (12.39) is replaced by a perimeter law in continuum Abelian gauge theories whereas in non-Abelian gauge theories the asymptotic freedom (11.68) leads to the preservation of the area law till the continuum limit. A lot of computer simulations with big volumes of the lattice and small δ support this picture. In principle, we do not need the gauge fixing on the lattice for the path integration. However, for a comparison with the continuum weak coupling limit the gauge fixing is necessary. The continuum limit of lattice gauge theories can be obtained and compared with continuum perturbative gauge theories in a formal perturbation expansion. For this purpose the lattice ξ-gauge is useful when we wish to compare lattice calculations with the ones in perturbative gauge theories (e.g., the renormalization and the short distance behavior [55]). In this context, the formal methods of Sect. 11.4 require a rigorous formulation. We can formulate Eq. (11.38) as

$$\delta^{-1} \sum_j \left(g_{<x,y>}^{-1} g_{<x,y+e_j>} - 1 \right) = C(x) \tag{12.40}$$

for each lattice point x. In Eq. (12.40), for typographical reasons we write $g_{xy} = g_{<x,y>}$, the sum is over the nearest neighbors. Then, the Faddeev-Popov identity (11.31) is mathematically well-defined and can be inserted in the expectation values.

A non-perurbative procedure of controlling the limit $\delta \to 0$ and the weak coupling relies on dividing the boxes of lengths δ into the ones of length $\frac{\delta}{n}$ and bounding the theory $\frac{\delta}{n}$ by the one with δ (block renormalization group developed by Balaban [26]). However, this ambitious program has not been carried out till the end. We may say that mathematically rigorous version of QCD and the Standard Model has not been established yet.

12.4 Exercises

12.1 Prove Eq. (12.6)
Hint: the simplest way is to express the Lagrangian in Fourier components as indicated below Eq. (12.7).
12.2 Calculate the lowest order integrals on the lattice in the perturbative Eq. (12.10). After calculation take the limit $\delta \to 0$.
12.3 By means of an expansion in V in Eq. (12.22) show that the limit $\delta \to 0$ of Eq. (12.22) gives the Feynman-Kac formula in the continuum.
12.4 Derive the polymer representation in the Abelian Higgs model using the expansion (12.36).
12.5 Apply the Faddeev-Popov procedure to the gauge fixing function of Eq. (12.40). Obtain the counterpart for the Faddeev-Popov determinant on the lattice.
12.6 Prove the area law in the non-Abelian case using an analog of Eq. (12.37) for group representations.

References

1. B.P. Abbott et al., Observation of gravitational waves from a binary black hole. Phys. Rev. Lett. **116**, 061102 (2016)
2. L.F. Abbott, The background method beyond one loop. Nucl. Phys. B **185**, 189 (1981)
3. L.F. Abbott, Introduction to the background field method. Acta Phys. Pol. B **13**, 33 (1982)
4. E.S. Abers, B.W. Lee, Gauge theories. Phys. Reports **C9**, 1 (1973)
5. L. Accardi, On the quantum Feynman-Kac formula. Rendiconti Seminario Mat. Fis. (Milano) **48**, 135 (1978)
6. L. Accardi, A. Frigerio, J.T. Lewis, Quantum stochastic processes Publ. RIMS, Kyoto Univ. **18**, 97(1982)
7. S. Albeverio, R. Hoegh-Krohn, Uniqueness of the physical vacuum and the Wightman functions in the infinite volume limit for some non-polynomial interactions. Commun. Math. Phys. **30**, 171 (1973)
8. S. Albeverio, R. Hoegh-Krohn, *Mathematical Theory of Feynman Path Integrals* (Springer, Berlin, 1976)
9. S. Albeverio, R. Hoegh-Krohn, Energy forms, Hamiltonians and distorted Brownian paths. J. Math. Phys. **18**, 907 (1977)
10. S. Albeverio, R. Hoegh-Krohn, Dirichlet forms and diffusion processes on rigged Hilbert spaces. Z. Wahrsch. verw. Geb. **40**, 1 (1977)
11. S. Albeverio, G. Gallavotti, R. Hoegh-Krohn, Some results for exponential interaction in two and more dimensions. Commun. Math. Phys. **70**, 187 (1979)
12. S. Albeverio, Ph. Blanchard, R. Hoegh-Krohn, Feynman path integrals and the trace formula for the Schrödinger operators. Commun. Math. Phys. **83**, 49 (1982)
13. S. Albeverio, Z. Brzezniak, Finite dimensional approximation approach to oscillatory integrals and stationary phase in infinite dimensions. J. Funct. Anal. **113**, 117 (1993)
14. S. Albevero, S. Kusuoka, A basic estimate for two-dimensional stochastic holonomy along Brownian bridges. J. Funct. Anal. **127**, 132 (1995)
15. S. Albeverio, Z. Brzezniak, Z. Haba, Schrödinger equation with potentials which are Laplace transforms of measures. Potential Anal. **9**, 65 (1998)
16. S. Albeverio, S. Mazzucchi, Feynman path integrals for the time dependent quartic oscillator. C.R. Acad. Sci. Paris, Ser. I **341**, 647 (2005)
17. S. Albeverio, S. Mazzucchi, The time dependent quartic oscillator- a Feynman path integral approach. J. Func. Anal. **238**, 471 (2006)

© The Editor(s) (if applicable) and The Author(s), under exclusive license to Springer   223
Nature Switzerland AG 2023
Z. Haba, *Lectures on Quantum Field Theory and Functional Integration*,
https://doi.org/10.1007/978-3-031-30712-6

18. A. Albrecht, P. Ferreira, M. Joyce, T. Prokopec, Inflation and squeezed quantum states. Phys. Rev. D **50**, 4807 (1994)
19. R. Alicki, M. Fannes, *Quantum Dynamical Systems* (Oxford University Press, 2001)
20. C. Anastopoulos, B.L. Hu, A master equation for gravitational decoherence:probing the textures of spacetime. Class. Quant. Gravity **30**, 165007 (2013)
21. I.Ya. Arefeva, A.A. Slavnov, L.D. Faddeev, Generating functional for the S-matrix in gauge-invariant theories. Theor. Math. Phys. **21**, 1165 (1974)
22. A. Ashtekar, A. Magnon, Quantum fields in curved space-time. Proc. R. Soc. Lond. **A346**, 375 (1975)
23. A. Ashtekar, J. Lewandowski, H. Sahlmann, Polymer and Fock representations for a scalar field. Class. Quant. Grav. **20**, L11 (2002)
24. A. Ashtekar, A. Corichi, A. Kesavan, Emergence of classical behavior in the early Universe. Phys. Rev. D **102**, 023512 (2020)
25. G. Baker, Self-interacting boson quantum field theory and the thermodynamic limit in d dimensions. J. Math. Phys. **16**, 1324 (1975)
26. T. Balaban, Large field renormalization. I. The basic step of the R operation. Commun. Math. Phys.**122**, 175 (1989)
27. G. Barton, Quantum mechanics of the inverted oscillator potential. Ann. Phys. (NY) **166**, 322 (1986)
28. A.O. Barut, R. Raczka, *Theory of Group Representations and Applications* (World Scientific, Singapore, 1986)
29. H. Bateman, *Higher Transcendental Functions*, vol. 2 (McGrow-Hill Book Company, 1953)
30. R. Benguria, M. Kac, Quantum Langevin equation. Phys. Rev. Lett. **46**, 1 (1981)
31. A.N. Bernal, M. Sanchez, Smoothness of time functions and the metric splitting of globally hyperbolic spacetimes. Commun. Math. Phys **257**, 43 (2005)
32. M.V. Berry, K.E. Mount, Semiclassical approximations in wave mechanics. Reports Prog. Phys. **35**, 315 (1972)
33. I. Bialynicki-Birula, Renormalization, diagrams, and gauge invariance. Phys. Rev. D **2**, 2877 (1970)
34. I. Bialynicki-Birula, J. Mycielski, Uncertainty relations for information entropy in wave mechanics. Commun. Math. Phys. **44**, 129 (1975)
35. N.D. Birrell, P.C.W. Davis, *Quantum Fields in Curved Space* (Cambridge University Press, Cambridge, 1982)
36. J.D. Bjorken, S.D. Drell, *Relativistic Quantum Fields* (McGraw-Hill, New York, 1965)
37. K. Bleuler, Eine neue Methode zur Behandlung der longitundinalen und skalaren Photonen. Helv. Phys. Acta **23**, 567 (1950)
38. S. Boughn, T. Rothman, Aspects of graviton detection:graviton emission and absorption by atomic hydrogen. Class. Quant. Grav. **23**, 5839 (2006)
39. M. Born, W. Heisenberg, P. Jordan, Zur Quantenmechanik. Zeit. Phys. **34**, 858 (1925)
40. R. Brandenberger, V. Mukhanov, T. Prokopec, Entropy of a classical stochastic field and cosmological perturbations. Phys. Rev. Lett. **69**, 3606 (1992)
41. R. Brandenberger, V. Mukhanov, T. Prokopec, Entropy of the gravitational field. Phys. Rev. D **48**, 2443 (1993)
42. J. Bros, U. Moschela, Two-point functions and quantum fields in de Sitter universe. Rev. Math. Phys. **8**, 327 (1996)
43. D. Brydges, J.Fröhlich, E. Seiler, On the construction of quantized gauge fields III. The two-dimensional Abelian Higgs model without cutoffs. Commun. Math. Phys. **79**, 353 (1981)
44. D. Brydges, P. Federbush, A lower bound for the mass of a random Gaussian lattice. Commun. Math. Phys. **62**, 79 (1982)
45. D. Brydges, J. Fröhlich, T. Spencer, The random walk representation of classical spin systems and correlation inequalities. Commun. Math. Phys. **83**, 123 (1982)
46. T.S. Bunch, P.C.W. Davis, Quantum field theory in de Sitter space: renormalization by point splitting. Proc. R. Soc. Lond. **360**, 117 (1978)

47. D. Burgarth, N. Galke, A. Hahn, L. van Luijk, State-dependent Trotter limits and their approximations. arXiv:2209.14787
48. A.O. Caldeira, A.J. Leggett, Path integral approach to quantum Brownian motion. Physica A **121**, 587 (1983)
49. H.B. Callen T.A. Welton, Irreversibility and generalized noise. Phys. Rev. **83**, 34 (1951)
50. R.H. Cameron, The Ilstow and Feynman integrals. J. d'Analyse Math. **10**, 287 (1962)
51. E. Carlen, P. Kree, L^p estimates on iterated stochastic integrals. Ann. Prob. **19**, 354 (1991)
52. S. Chandrasekhar, Stochastic problems in physics and astronomy. Rev. Mod. Phys. **15**, 1 (1943)
53. T.P. Cheng, L.F. Li, *Gauge Theory and Elementary Particle Physics* (Clarendon Press, Oxford, 1984)
54. Y. Choquet-Bruhat, Probleme de Cauchy pour le systeme integro-differentielle d'Einstein-Liouville. Ann. l'Inst. Fourier **3**, 181 (1971)
55. K. Cichy, K. Jansen, P. Korcyl, Non-perturbative renormalization in coordinate space for $N_f = 2$ maximally twisted mass fermions with tree-level Symanzik improved gauge action. Nucl. Phys. B **865**, 268 (2012)
56. S. Coleman, Laws of hadronic matter, in *Proceeding of the 11th Course of "Ettore Majorana"*, ed. by A. Zichichi (Academic, New York, 1975)
57. A. Corichi, J. Cortez, H. Quevedo, Schrödinger representation for a scalar field on curved spacetime. Phys. Rev. D **66**, 085025 (2002)
58. P. Courrage, P. Priouret, P. Renouard, M. Yor, Oscillateur anharmonique, processus de diffusion et measures quasi-invariantes, Asterisque. Tome 22–29 (1975)
59. M.P. Dabrowski, T. Stachowiak, M. Szydlowski, Phantom cosmologies. Phys. Rev. D **68**, 103519 (2003)
60. Yu.L. Daletskii, Functional integrals connected with operator evolutionary equations. Russ. Math. Surv. **22**, 1 (1967)
61. G.F. de Angelis, D. de Falco, G. Di Genova, Random fields on Riemannian manifolds: a constructive approach. Commun. Math. Phys. **103**, 297 (1986)
62. D. Deutsch, Uncertainty in quantum measurements. Phys. Rev. Lett. **50**, 631 (1983)
63. C. Deutsch, M. Lavaud, Equilibrium properties of a two-dimensional Coulomb gas. Phys. Rev. A **9**, 2598 (1974)
64. L. Di Luzio, M. Giannotti, E. Nardi, L. Visinnelli, The landscape of QCD axion models. Phys. Reports **870**, 1 (2020). arxiv:2003.01100 [hep-ph]
65. J. Dimock, Markov quantum fields on a manifold. Rev. Math. Phys. **16**, 243 (2004)
66. J. Dimock, Algebras of local observables on a manifold. Commun. Math. Phys. **77**, 219 (1980)
67. P.A.M. Dirac, The fundamental equations of quantum mechanics. Proc. R. Soc. (London) **109**, 642 (1925)
68. P.A.M. Dirac, Classical theory of radiating electrons. Proc. R. Soc. (London) **A167**, 148 (1938)
69. L. Dolan, R. Jackiw, Functional evaluation of the effective potential. Phys. Rev. D **9**, 3320 (1974)
70. J.D. Dollard, Ch.N. Friedman, On strong product integration. J. Funct. Anal. **28**, 309 (1978)
71. H. Doss, On a stochastic solution of the Schroedinger equation with analytic coefficients. Commun. Math. Phys. **73**, 247 (1980)
72. H. Doss, On a probabilistic approach to the Schrödinger equation with a time-dependent potential. J. Funct. Anal. **260**, 1824 (2011)
73. L. Durand, E. Mendel, Field-strength formulation of gauge theories: Transformation of the functional integral. Phys. Rev. D **26**, 1368 (1982)
74. B. Durhus, J. Fröhlich, A connection between ν-dimensional Yang-Mills theory and $(\nu - 1)$-dimensional non-linear σ-models. Commun. Math. Phys. **75**, 103 (1980)
75. A. Einstein, L. Hopf, Statistische Untersuchung der Bewegung eines Resonators in einem Strahlungsfeld. Ann. d. Phys. **33**, 1105 (1910)
76. D. Elworthy, A. Truman, Feynman maps, Cameron-Martin formulae and anharmonic oscillators. Ann. Inst. Henri Poincare **41**, 115 (1984)

77. H. Euler, W. Heisenberg, Folgerungen aus der diracschen theorie des positrons. Z. Phys. **98**, 714 (1936)
78. H. Ezawa, J.R. Klauder, L.A. Shepp, Path space picture for Feynman-Kac averages. Ann. Phys. **88**, 588 (1974)
79. L.D. Faddeev, V.N. Popov, Feynman diagrams for the Yang-Mills field. Phys. Lett. **25B**, 30 (1967)
80. L.D. Faddeev, A.A. Slavnov, *Gauge Fields. Introduction to Quantum Theory*, 2nd edn. (CRC Press, 1993)
81. F. Falceto, Canonical quantization of the electromagnetic field in arbitrary ξ -gauge. hep-th arXiv:2211.16870
82. M. Falconi, Cylindrical Wigner measures. Documenta Math. **23**, 1677 (2018)
83. X. Feal, A. Tarasov, R. Venugopalan, QED as a many-body theory of worldlines: I. General formalism and infrared structure. arXiv:2206.04188
84. J. Feldbrugge, N. Turok, Existence of real time quantum path integrals. arXiv:2207.12798
85. J. Feldman, A relativistic Feynman-Kac formula. Nucl. Phys. **52**, 608 (1973)
86. R.P. Feynman, F.L. Vernon, The theory of general quantum system interacting with a linear dissipative system. Ann. Phys. **24**, 118 (1963)
87. R.P. Feynman, A.R. Hibbs, *Qantum Mechanics and Path Integrals* (McGrawHill, New York, 1965)
88. R. Floreanini, C.T. Hill, R. Jackiw, Functional representation for the isometries of de Sitter space. Ann. Phys. **175**, 345 (1987)
89. G.B. Folland, *Harmonic Analysis in Phase Space* (Princeton University Press, Princeton, 1989)
90. G.W. Ford, M. Kac, On the quantum Langevin equation. J. Stat. Phys. **46**, 803 (1987)
91. G.W. Ford, J.T. Lewis, R.F. O'Connell, Quantum Langevin equation. Phys. Rev. A **37**, 4419 (1988)
92. G.W. Ford, J.T. Lewis, R.F. O'Connell, Magnetic-field effects on the motion of a charged particle in a heat bath. Phys. Rev. A **41**, 5287 (1990)
93. J. Fröhlich, Classical and quantum statistical mechanics in one and two dimensions: two-component Yukawa and Coulomb systems. Commun. Math. Phys. **47**, 233 (1976)
94. J. Fröhlich, New super-selection sectors ("soliton-states") in two dimensional Bose quantum field models. Commun. Math. Phys. **47**, 269 (1976)
95. J. Fröhlich, On the triviality of $\lambda\phi$ d4 theories and the approach to the critical point in $d > 4$ dimensions. Nucl. Phys. **200**[FS4], 281 (1982)
96. M. Freidlin, *Functional Integration and Partial Differential Equations* (Princeton University, Press, 1995)
97. M.P. Fry, Nonperturbative quantization of the electroweak models, electrodynamic sector. Phys. Rev. D **91**, 085026 (2015)
98. D. Fujiwara, A construction of the fundamental solution for the Schrödinger equation. J. d'Analyse Math. **35**, 41 (1979)
99. S.A. Fulling, *Aspects of Quantum Field Theory in Curved Spacetime* (Cambridge University Press, Cambridge, 1989)
100. S. Gao, Lindblad approach to quantum dynamics of open systems. Phys. Rev. B **57**, 4509 (1998)
101. C. Gardiner, *Quantum Noise* (Springer, New York, 2004)
102. S. Gasiorowicz, *Elementary Particle Physics* (Wiley, New York, 1966)
103. I.M. Gelfand, G.E. Shilov, *Generalized Functions*, vol. 1 (AMS, Providence, Rhode Island, 1964)
104. I.M. Gelfand, N.. Ya.. Vilenkin, *Generalized Functions*, vol. 4 (AMS Chelsea Publishing, New York, 1964)
105. I.M. Gelfand, A.M. Yaglom, Integration in functional spaces and its applications in quantum physics. J. Math. Phys. **1**, 48 (1960)
106. M. Gell-Mann, F. Low, Bound states in quantum field theory. Phys. Rev. **84**, 350 (1951)

107. G.W. Gibbons, S.W. Hawking, Action integrals and partition functions in quantum gravity. Phys. Rev. D **15**, 2738 (1977)
108. I.I. Gikhman, A.V. Skorohod, *Stochastic Differential Equations* (Springer, New York, 1972)
109. J. Ginibre, in *Statistical Mechanics and Quantum Field Theory*, eds. by C. de Witt, R. Stora (Gordon and Breach, New York, 1971)
110. R.J. Glauber, The quantum theory of optical coherence. Phys. Rev. **130**, 2529 (1963)
111. J. Glimm, A. Jaffe, *Quantum Physics. The Functional Integral Point of View* (Springer, New York, 1981)
112. J. Glimm, A. Jaffe, Boson quantum field models, in *Mathematics of Contemporary Physics*. ed. by R.F. Streater (Academic, New York, 1972)
113. S. Goldstein, J.L. Lebowitz, On the (Boltzmann) entropy of non-equilibrium systems. Physica D **193**, 53 (2004)
114. V. Gorini, A. Kossakowski, E.C.G. Sudarshan, Completely positive dynamical semigroups of N-level systems. J. Math. Phys. **17**, 821 (1976)
115. I.S. Gradshteyn, I.M. Ryzhik, *Tables of Integrals Series and Products* (Academic, New York, 1965)
116. O.W. Greenberg, Haag's theorem and clothed operators. Phys. Rev. **115**, 706 (1959)
117. V.N. Gribov, Quantization of non-Abelian gauge theories. Nucl. Phys. B **139**, 1 (1978)
118. D.J. Griffiths, *Introduction to Electrodynamics*, 3rd edn. (Prentice-Hall Inc., New Jersey, 1999)
119. L.P. Grishchuk, Y.V. Sidorov, Squeezed quantum states of relic gravitons and primordial density fluctuations. Phys. Rev. D **42**, 3413 (1990)
120. D.J. Gross, F. Wilczek, Ultraviolet behavior of non-abelian gauge theories. Phys. Rev. Lett. **30**, 1343 (1973)
121. D.J. Gross, R.D. Pisarski, L.G. Yaffe, QCD and instantons at finite temperature. Rev. Mod. Phys. **53**, 43 (1981)
122. L. Gross, Potential theory on Hilbert space. J. Funct. Anal. **1**, 123 (1967)
123. M. Grothaus, M.J. Oliveira, J.L. da Silva, L. Streit, Self-avoiding fractional Brownian motion-the Edwards model. J. Stat. Phys. **145**, 1513 (2011)
124. F. Guerra, L. Rosen, B. Simon, The $P(\phi)_2$ Euclidean quantum field theory as classical statistical mechanics. Ann. Math. **101**, 111 (1975)
125. F. Guerra, L. Rosen, B. Simon, Boundary conditions for the euclidean field theory. Ann. Inst. Henri Poincare **15**, 231 (1976)
126. F. Guerra, L. Rosen, B. Simon, The vacuum energy for $P(\phi)_2$: Infinite volume limit and coupling constant dependence. Commun. Math. Phys. **29**, 233 (1973)
127. V. Guillemin, S. Sternberg, *Geometric Asymptotics* (American Mathematical Society, Providence, 1977)
128. A. Guth, S.-Y. Pi, Quantum mechanics of the scalar field in the new inflationary universe. Phys. Rev. D **32**, 679 (1985)
129. S.N. Gupta, Theory of longitudinal photons in quantum electrodynamics. Proc. R. Soc. (London)**63**, 681 (1950)
130. J. Guven, B. Liebermann, Ch.T. Hill, Schrödinger-picture field theory in Robertson-Walker flat spacetimes. Phys. Rev. D **39**, 438 (1989)
131. R. Haag, On quantum field theories. Dan. Mat, Fys. Medd. **29**, 1 (1955)
132. R. Haag, *Local Quantum Physics. Fields, Particles, Algebras* (Springer, Berlin, 1992)
133. Z. Haba, Some non-markovian Osterwalder-Schrader fields. Ann. Inst. Henri Poincare **32**, 185 (1980)
134. Z. Haba, Behavior in strong fields of Euclidean gauge theories. Phys. Rev. D **26**, 3506 (1982)
135. Z. Haba, Behavior in strong fields of Euclidean gauge theories II. Phys. Rev. D **29**, 1716 (1984)
136. Z. Haba, Reflection positivity and quantum fields on a Riemann surface, in *Stochastic Processes, Physics and Mathematics*, eds. by S. Albeverio, G. Casati U. Cattaneo, D. Merlini, R. Moresi (World Scientific, Singapore, 1990)
137. Z. Haba, Stochastic interpretation of Feynman path integral. J. Math. Phys. **35**, 6344 (1994)

138. Z. Haba, Semiclassical stochastic representation of the Feynman integral. J. Phys. **A27**, 6457 (1994)

139. Z. Haba, Feynman integral and complex classical trajectories. Lett. Math. Phys. **37**, 223 (1996)

140. Z. Haba, Feynman integral in regularized nonrelativistic quantum electrodynamics. J. Math. Phys. **39**, 1766 (1998)

141. Z. Haba, *Feynman Integral and Random Dynamics in Quantum Physics* (Kluwer/Springer, Dordrecht, 1999)

142. Z. Haba, H. Kleinert, Master and Langevin equations for electromagnetic dissipation and decoherence of density matrices. Eur. Phys. J. B **21**, 553 (2001)

143. Z. Haba, H. Kleinert, Quantum Liouville and Langevin equations for gravitational radiation damping. Int. J. Mod. Phys. A **17**, 3729 (2002). arXiv:quant-ph/0101006

144. Z. Haba, State-dependent graviton noise in the equation of geodesic deviation. Eur. Phys. J. C **81**, 40 (2021)

145. Z. Haba, Graviton noise: the Heisenberg picture. Int. J. Mod. Phys. D32, 2350005 (2023). https://doi.org/10.1142/S0218271823500050

146. Z. Haba, Feynman-Kac path integral expansion around the upside-down oscillator. (2023) (to appear)

147. S.W. Hawking, G.F.R. Ellis, *The Large Scale Structure of Space Time* (Cambridge University Press, Cambridge, 1973)

148. S.W. Hawking, Zeta function regularization of path integrals in curved space. Commun. Math. Phys. **55**, 133 (1977)

149. G.C. Hegerfeldt, From Euclidean to relativistic fields and on the notion of Markoff fields. Commun. Math. Phys. **35**, 155 (1974)

150. W. Heisenberg, W. Pauli, Zur quantentheorie der Wellenfelder. II Zeit. Phys. **56**, 1 (1929); **59**, 168 (1930)

151. W. Heitler, *The Quantum Theory of Radiation* (Oxford Clarendon Press, 1954)

152. Ch. Henry, R.F. Kazarinov, Quantum noise in photonics. Rev. Mod. Phys. **68**, 801 (1996)

153. P.W. Higgs, Broken symmetries and the masses of gauge bosons. Phys. Rev. Lett. **13**, 508 (1964)

154. L. Hörmander, Fourier integral operators. I, Acta Math. **127**, 79 (1971)

155. J. Howland, Stationary scattering theory for time-dependent Hamiltonians. Math. Ann. **207**, 315 (1974)

156. B.L. Hu, J.P. Paz, Y. Zhang, Quantum Brownain motion in a general environment: Exact master equation with nonlocal dissipation and colored noise. Phys. Rev. D **45**, 2843 (1992)

157. R.L. Hudson, When is the Wigner quasi-probability density non-negative? Reports Math. Phys. **6**, 249 (1974)

158. N. Ikeda, S. Watanabe, *Stochastic Differential Equations and Diffusion Processes* (North Holland, 1981)

159. J. Iliopoulos, C. Itzykson, A. Martin, Functional methods and perturbation theory. Rev. Mod. Phys. **47**, 165 (1975)

160. K. Ito, Genaralized uniform complex measures in the Hilbertian metric space with their application to the Feynman path integral, in Proceedings of Fifth Berkley Sympoisum (University of California Press, Berkeley, 1967)

161. R. Jackiw, in *Field Theory and Particle Physics*, ed. by O. Eboli (World Scientific, Singapore, 1990)

162. J.D. Jackson, *Classical Electrodynamics* (Wiley, New York, 2021)

163. A. Jaffe, C.H. Taubes, *Vortices and Monopoles* (Birkhhäuser, Boston, 1980)

164. A. Jaffe, G. Ritter, Quantum field theory on curved backgrounds. II. Spacetime symmetries. Commun. Math. Phys. **270**, 545 (2007)

165. G. Jona-Lasinio, F. Martinelli, E. Scoppola, New approach to the semiclassical limit of quantum mechanics. Commun. Math. Phys. **80**, 223 (1981)

166. P. Jordan, W. Pauli, Zur Quantenelektrodynamik ladungsfreier Felder. Zeit. Phys. **47**, 151 (1928)

167. S. Kanno, J. Soda, J. Tokuda, Noise and decoherence induced by gravitons. Phys. Rev. D **103**, 044017 (2021)
168. J.I. Kapusta, Ch. Gale, *Finite Temperature Field Theory. Principles and Applications* (Cambridge University Press, Cambridge, 2006)
169. R.L. Karp, F. Mansouri, Product integral formalism and non-Abelian Stokes theorem. J. Math. Phys. **40**, 6033 (1999)
170. H. Kleinert, *Path Integrals*, 5th edn. (World Scientific, 2009)
171. R. Kubo, Statistical-mechanical theory of irreversible processes. I. general theory and simple applications to magnetic and conduction problems. J. Phys. Soc. Jpn **12**, 570 (1957)
172. H.-H. Kuo, *Gaussian Measures in Banach Spaces* (Springer, Berlin, 1975)
173. W.E. Lamb, R.C. Retherford, Fine strncture of the hydrogen atom by a microwave method. Phys. Rev. **72**, 241 (1947)
174. M. le Bellac, *A Short Introduction to Quantum Information and Quantum Computation* (Cambridge University Press, 2006)
175. H. Lehmann, K. Symanzik, W. Zimmermann, On the formulation of quantized field theories. Nuovo Cimento **1**, 205 (1955)
176. G. Lindblad, On the generators of quantum dynamical semigroups. Commun. Math. Phys. **48**, 119 (1976)
177. L.N. Lipatov, Divergence of the perturbation theory series and the quasiclassical theory. Sov. Phys. JETP **45**, 216 (1977)
178. D.V. Long, G.M. Shore, The Schrödinger wave functional and vacuum states in curved spacetime. Nucl. Phys. B **530**, 247 (1998)
179. J. Lörinczi, F. Hiroshima, V. Betz, *Feynman-Kac Type Theorems and Gibbs Meaures on Path Space* (Walter de Gruyter GmbH, Berlin, 2011)
180. M. Maggiore, *Gravitational Waves*, vol. 1 (Oxford University Press, 2007)
181. L. Mandel, E. Wolf, *Optical Coherence and Quantum Optics* (Cambridge University Press, 1995)
182. S. Mandelstam, Quantum electrodynamics without potentials, Ann. Phys. (NY) **19**, 1 (1962)
183. P.C. Martin, J. Schwinger, Theory of many-particle systems. I Phys. Rev. **115**, 1342 (1959)
184. V.P. Maslov, M.V. Fedoriuk, *Semiclassical Approximation in Quantum Mechanics* (D. Reidel, Dordrecht, Holland, 1981)
185. K. Maurin, *Methods of Hilbert Spaces* (PWN, Warszawa, 1967)
186. S. Mazzucchi, Feynman path integrals for the inverse quartic oscillator. J. Math. Phys. **49**, 093502 (2008)
187. E. Merzbacher, *Quantum Mechanics*, 3rd edn (Wiley, 1998)
188. P.W. Milonni, *The Quantum Vacuum. An Introduction to Quantum Electrodynamics* (Academic, New York, 1994)
189. C.B. Morrey, *Multiple Integrals in the Calculus of Variations* (Springer, Berlin, 1966)
190. E. Nelson, Feynman integrals and the Schrödinger equation. J. Math. Phys. **5**, 332 (1964)
191. K. Osterwalder, R. Schrader, Axioms for Euclidean Green's functions. Commun. Math. Phys. **31**, 83 (1973); corr. **42**, 281 (1975)
192. K. Osterwalder, E. Seiler, Gauge field theories on a lattice. Ann. Phys. **110**, 440 (1978)
193. M. Parikh, F. Wilczek, G. Zahariade, The noise of gravitons. Int. J. Mod. Phys **D29**, 2042001(2020). arXiv:2005.07211 [hep-th]
194. M. Parikh, F. Wilczek, G. Zahariade, Signatures of the quantization of gravity at gravitational wave detectors. Phys. Rev. D **104**, 046021 (2021)
195. P.J.E. Peebles, B. Ratra, The cosmologial constant and dark energy. Rev. Mod. Phys. **75**, 559 (2003)
196. M.E. Peskin, D.V. Schroeder, *An Introduction to Quantum Field Theory* (Perseus, Reading MA, 1995)
197. H.D. Politzer, Reliable perturbative results for strong interactions? Phys. Rev. Lett. **30**, 1346 (1973)
198. A.M. Polyakov, Compact gauge fields and the infrared catastrophe. Phys. Lett. B **59**, 82 (1975)
199. A.M. Polyakov, Quantum geometry of bosonic strings. Phys. Lett. B **103**, 207 (1981)

200. J. Preskill, M.B. Wise, F. Wilczek, Cosmology of the invisible axion. Phys. Lett. **120B**, 127 (1983)

201. T.C. Quinn, R.M. Wald, Axiomatic approach to electromagnetic and gravitational radiation reaction of particles in curved space-time. Phys. Rev. D **56**, 3381 (1997)

202. R. Rajaraman, Some non-perturbative semi-classical methods in quantum field theory (a pedagogical review). Phys. Rep. **C21**, 227 (1975)

203. M. Reed, B. Simon, *Methods of Modern Mathematical Physics*, vol. 1 (Academic, New York, 1975)

204. M. Reed, B. Simon, *Methods of Modern Mathematical Physics*, vol. 2 (Academic, New York, 1975)

205. W.P. Reinhardt, Complex coordinates in the theory of atomic and molecular structure and dynamics. Ann. Rev. Phys. Chem. **33**, 233 (1982)

206. H. Risken, *The Fokker-Planck Equation* (Springer, Berlin, 1989)

207. F. Rohrlich, *Classical Charged Particles. Foundations of Their Theory* (Addison-Wesley, Reading, 1965)

208. W. Rudin, *Real and Complex Analysis* (McGraw-Hill Inc., 1974)

209. L.H. Ryder, *Quantum Field Theory*, 2nd edn. (Cambridge University Press, 1996)

210. M. Salmhofer, *Renormalization: An Introduction* (Springer, New York, 1998)

211. R. Schrader, Towards a constructive approach of a gauge invariant massive $P(\phi)_2$ theory. Commun. Math. Phys. **58**, 299 (1978)

212. Ch. Schubert, An introduction to the worldline technique for quantum field theory calculations. Acta Phys. Polon. B **27**, 3965 (1995)

213. B.F. Schutz, *A First Course in General Relativity* (Cambridge University Press, Cambridge, 1985)

214. S.S. Schweber, *An Introduction to Relativistic Quantum Field Theory* (Row Peterson and Co., Evanstone, Ill., 1961)

215. J. Schwinger, On gauge invariance and vacuum polarization. Phys. Rev. **82**, 664 (1951)

216. E. Seiler, Schwinger functions for the Yukawa model in two dimensions with space-time cutoff. Commun. Math. Phys. **42**, 163 (1975)

217. E. Seiler, in *Proceeding of Poiana Brasov Summer School. Progress in Physics*, vol. 5, eds. by P. Dita, V. Georgescu, R. Purice (Birkhäuser, Basel, 1982), p. 263

218. E. Seiler, *Gauge Theories as a Problem of Constructive Field Theory and Statistical Physics LNPh*, vol. 159 (Springer, 1982)

219. E. Seiler, I.-O. Stamatescu, A note on the loop formula for the fermionic determinant. J. Phys. **A49**, 335401 (2016)

220. M. Silverman, *More Than One Mystery. Explorations in Quantum Interference* (Springer, New York, 1995)

221. B. Simon, *The $P(\phi)_2$ Euclidean Quantum Field Theory* (Princeton University Press, Princeton, 1974)

222. B. Simon, Notes on infinite determinants of Hilbert space operators. Adv. Math. **24**, 244 (1977)

223. B. Simon, Resonances and complex scaling: a rigorous overview. Int. J. Quantum Chem. **14**, 529 (1978)

224. B. Simon, *Functional Integration and Quantum Physics* (Academic, New York, 1981)

225. A.V. Skorohod, *Integration in Hilbert Space* (Springer, Berlin, 1974)

226. A.V. Skorohod, *Random Processes with Independent Increments* (Springer, Dordrecht, 1991)

227. H. Spohn, Entropy production for quantum dynamical semigroups. J. Math. Phys. **19**, 1227 (1978)

228. R.F. Streater, A.S. Wightman, *PCT, Spin and Statistics and All That* (W.B. Benjamin, New York, 1968)

229. K. Symanzik, in *Local Quantum Theory*, ed. by R. Jost (Academic, 1969)

230. J. Teschner, Liouville theory revisited. Class. Quant. Gravity **18**, R153 (2001)

231. M.C. Teich, B.E.A. Saleh, Squeezed states of light. Quantum Opt. **1**, 153 (1989)

232. G.t'Hooft, Renormalization of massless Yang-Mills fields. Nucl. Phys. **B33**, 173 (1971)

233. G.t'Hooft, M. Veltman, Regularization and renormalization of gauge fields. Nucl. Phys. **B44**, 189 (1972)

234. G.t'Hooft, in *Acta Universitatis Wratislavensis*, vol. 38 (1976), ed. by B. Jancewicz, XII Winter School in Karpacz, 1975

235. H. F. Trotter, On the product of semi-groups of operators, Proc.Amer.Math.Soc.**10**,545(1959)

236. W.G. Unruh, Notes on black-hole evaporation. Phys. Rev. D **14**, 870 (1976)

237. S. Vagnozzi, A. Loeb, The challenge of ruling out inflation via the primordial gravitational background. Astroph. J. Lett. **939**, L22 (2022)

238. P. Nieuvenhuizen, Supergravity. Phys. Rep. **32C**, 250 (1977)

239. R.M. Wald, *Quantum Field Theory in Curved Spacetime and Black Hole Thermodynamics* (The University of Chicago Press, Chicago, 1994)

240. S. Weinberg, *The Quantum Theory of Fields*, vol. 1 (The Press Sindicate of the University of Cambridge, Foundations, 1995)

241. M.J. Westwater, On Edwards model for long polymer chains. Comm. Math. Phys. **72**, 131 (1980)

242. E.P. Wigner, On the quantum correction for thermodynamic equilibrium. Phys. Rev. **40**, 749 (1932)

243. E.P. Wigner, On unitary representations of the inhomogeneous Lorentz group. Ann. Math. **40**, 149 (1939)

244. F. Wilczek, Axions and family symmetry breaking. Phys. Rev. Lett. **49**, 1549 (1982)

245. K.G. Wilson, Confinement of quarks. Phys. Rev. D **10**, 2445 (1974)

246. C.N. Yang, D. Feldman, The S-matrix in the Heisenberg representation. Phys. Rev. **79**, 972 (1950)

247. C.N. Yang, R.L. Mills, Conservation of isotopic spin and isotopic gauge invariance. Phys. Rev. **96**, 191 (1954)

248. H. Yukawa, On the interaction of elementary particles. I.,Proceed.Phys.-Math.Soc.Japan,**17**,48(1935)

249. W.H. Zurek, Decoherence, einselection and the quantum origin of classical. Rev. Mod. Phys. **75**, 715 (2003)

Index

© The Editor(s) (if applicable) and The Author(s), under exclusive license to Springer   233
Nature Switzerland AG 2023
Z. Haba, *Lectures on Quantum Field Theory and Functional Integration*,
https://doi.org/10.1007/978-3-031-30712-6

Index   235

Printed in the United States
by Baker & Taylor Publisher Services

Printed in the United States
by Baker & Taylor Publisher Services